Springer Undergraduate Mathematics Series

For other titles published in this series, go to
www.springer.com/series/3423

Martin D. Crossley

Essential Topology

 Springer

Dr. Martin D. Crossley
University of Wales Swansea
Department of Mathematics
Singleton Park
SA2 8PP Swansea, Wales
United Kingdom
m.d.crossley@swansea.ac.uk

ISSN 1615-2085
ISBN 978-1-85233-782-7   ISBN 978-1-84628-194-5 (eBook)
DOI 10.1007/978-1-84628-194-5
Springer London Dordrecht Heidelberg New York

British Library Cataloguing in Publication Data
A catalogue record for this book is available from the British Library

Library of Congress Control Number: 2010931641

Mathematics Classification Codes: 54-01, 55-01, 57-01, 54A05, 54B05, 54B10, 54B15, 54C05, 54D05, 54D10, 54D30, 55M25, 55N10, 55P35, 55P40, 55Q05, 55R10, 55U10, 55U20, 57R05, 55P10, 55P20, 55Q10, 55U15, 57M05

Printed on acid-free paper

Springer is part of Springer Science+Business Media (www.springer.com)

Preface

This book grew out of courses given at Swansea University to second- and third-year undergraduates. It is designed to provide enough material for a one-year course and splits naturally into a preliminary topology course (Chapters 2–6) and a follow-on course in algebraic topology (Chapters 7–11).

It is often said that topology is a subject which is poorly served for textbooks, and when preparing the lecture courses I found no book that was both accessible to our undergraduates and relevant to current research in the field. This book is an attempt to fill that gap. It is generally accepted that a one-year course on topology is not long enough to take a student to a level where she or he can begin to do research, but I have tried to achieve that as nearly as possible. By omitting some of the more traditional material such as metric spaces, this book takes a student from a discussion of continuity, through a study of some topological properties and constructions, to homotopy and homotopy groups, to simplicial and singular homology and finally to an introduction to fibre bundles with a view towards K-theory. These are subjects which are essential for research in algebraic topology, and desirable for students pursuing research in any branch of mathematics. In fact, if I may be so bold as to say so, the subjects covered by this book are those areas of topology which *all* mathematics undergraduates should ideally see. In that sense, the material is *essential* topology.

With this range of topics, and the low starting level, the coverage of each subject is, inevitably, not exhaustive. For example, there are many results about connectivity whose proofs could be understood by undergraduates at this level, but which do not appear in this book. Instead, a representative sample of such results is included, together with enough examples that the reader should fully understand the results presented. In an undergraduate course it seems better to present a brief account of several topics and give a feel for the overall shape of a subject, rather than an in-depth study of a small number of topics.

Some of the deeper results included are presented without proof, so that the student may meet an important theorem in the area even though the proof would lengthen the book unacceptably. In every such case references are given to books which do contain a complete proof.

Given the target audience, the book is designed to require as little prior knowledge as possible. Anyone who has some basic familiarity with functions, such as from a beginning course on calculus, should be able to follow the first four chapters. From Chapter 5 onwards, a little knowledge of algebra is required, in particular equivalence relations for Chapters 5 and 6, some familiarity with groups for Chapters 8 to 11, and with linear algebra and quotient groups for Chapters 9 and 10.

There is a short bibliography included, listing books where students can find details of the proofs which have been omitted. I have not included a list of further reading, as there are many books in topology and algebraic topology that should be intelligible to someone who has read through this book. The choice of which follow-on text to use is a matter of personal taste or, for students embarking on postgraduate study, is something that their supervisor will advise them about.

Acknowledgements

In writing on topology I must first thank John Hubbuck and Michael Crabb who taught me the subject as an undergraduate, and as a postgraduate.

I would also like to express my gratitude to my colleagues here at Swansea, particularly Francis Clarke who has passed on many helpful observations on communicating topology, and Geoff Wood who thought of Example 5.23. Thanks are also due to Simon Cowell and Nikki Burt who read preliminary versions of the manuscript and offered a number of helpful comments.

In the preparation of this book Karen Borthwick, Jenny Wolkowicki and Frank Ganz of Springer Verlag have been incredibly helpful, patiently leading me through all the different stages of the publishing process, for which I am very grateful. I also greatly appreciate the work of the anonymous reviewers who provided a wealth of helpful and perceptive comments. Since publication many readers have kindly pointed out errors. I am deeply grateful to all, but must single out Nikolaus Vonessen for his extensive list.

Finally I wish to thank my family: Ailsa, Calum and Jenny, for being a constant source of encouragement and distraction.

Contents

1
Introduction

Topology is one of the better-known areas of modern mathematics. Most people have heard the statement that a topologist is someone who cannot tell the difference between a tea cup and a doughnut. This is true, and we will see why in Chapter 5. Clearly, then, topology ignores some things and perceives similarities between apparently dissimilar objects. As we will discover, the key to what topology ignores, and to what it concentrates on, is the behaviour of continuous functions. Topology studies the ways in which the properties of the domain and range determine the behaviour of a continuous function.

For example, every continuous integer-valued function on the real line, $\mathbf{R}$, is constant. We do not need to know anything about the function apart from the fact that it is continuous and its range is $\mathbf{Z}$, the set of integers. Then, somehow, the natures of $\mathbf{R}$ and $\mathbf{Z}$ force that function to be constant. What matters here is the "topology" of $\mathbf{R}$ and $\mathbf{Z}$.

Another example is the fact that any continuous real function defined on the interval $[0, 1]$ (of all real numbers x such that $0 \leq x \leq 1$) must be bounded, i.e., there are some numbers j, k such that $j < f(x) < k$ for all $x \in [0, 1]$. By contrast, a continuous function defined on the open interval $(0, 1)$ (of all real numbers x such that $0 < x < 1$) need not be bounded. The function $f(x) = 1/x$ is an example of an unbounded function on $(0, 1)$. So something about $[0, 1]$ causes functions to be bounded whereas $(0, 1)$ does not. Again, it is the topology of $[0, 1]$ and the topology of $(0, 1)$ which produce this different behaviour.

Finally, in complex analysis there is a very clear example of topology in action. If you have a disc-like region, and a complex-valued function defined

M.D. Crossley, *Essential Topology*, Springer Undergraduate
Mathematics Series, DOI 10.1007/978-1-84628-194-5_1,
© Springer-Verlag London Limited 2010

everywhere on that region (i.e., without any poles in the region) then every contour integral of the function in that region will be 0. If, on the other hand, you take an annulus (the region between two concentric circles) and a complex-valued function defined everywhere on that region, then it does not follow that every contour integral will be 0. For example, the function may extend to one defined on the whole complex plane, but with poles inside the smaller circle. In that case a contour that loops round that region will give a non-zero integral.

An annulus with a "non-trivial" contour

In order to study how the domain and range affect the behaviour of continuous functions, we first need to clarify exactly what we mean by continuity, which we do in Chapter 2. We will then go on to generalize this notion in Chapter 3, before beginning our study of topology properly in Chapter 4.

2
Continuous Functions

As topology is essentially just the study of continuous functions, we should start by clarifying exactly what we mean by the word "continuous". We will begin by considering the most familiar type of function, namely those which are defined for some (or all) real numbers, and which return a real number, i.e., functions $f : S \to \mathbf{R}$ for some subset $S \subset \mathbf{R}$.

2.1 Naïve Continuity

Perhaps the simplest way to say that such a function is continuous would be to say that one can draw its graph without taking the pencil off the paper. For example, a function whose graph looks like

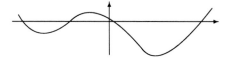

would be continuous in this sense. But if we look at the function $f(x) = 1/x$, then we see that things are not so simple. The graph of this has two parts - one part corresponding to negative x values, and the other to positive x values. The function is not defined at 0, so we certainly can't draw both parts of this graph without taking our pencil off the paper.

However, if we pick any point a in the domain of f, and look at the graph of f near a, and magnify this enough (i.e., look at a small enough area around a),

M.D. Crossley, *Essential Topology*, Springer Undergraduate
Mathematics Series, DOI 10.1007/978-1-84628-194-5_2,
© Springer-Verlag London Limited 2010

then we'll just get a line. For example, around the point $a = 0.3$, the graph looks like

Graph of $f(x) = 1/x$ near the point $x = 0.3$

Of course, this portion can be drawn in one movement of the pencil. And any point in the domain of $f(x) = 1/x$ has this property: The function is continuous near every point in its domain. Such a function deserves to be called continuous. So this characterization of continuity in terms of graph-sketching is too simplistic.

Instead, we must look at a function locally. If we can agree how to define what is meant by saying "f is continuous at the point a (in the domain of f)", then we can simply say that f is continuous if it is continuous at every point in its domain.

We can again use the graph to define what it means for a function f to be continuous at a single point a. It should mean that if you draw the graph of f, omitting the point $f(a)$, then you can predict the value of $f(a)$, based on the value of f at the points around a. For example, if, when we omit one point, the graph of f looks like this:

then we can confidently guess what the value of f is at the missing point, because we expect the graph to look like

But, of course, f needn't behave as we expect. There is nothing to stop the graph of f from looking like this:

For example, the function f defined by

$$f(x) = \begin{cases} x^2 & \text{if } x \neq 0, \\ 2 & \text{if } x = 0 \end{cases}$$

has such a graph.

In that case we would want to say that the function is discontinuous. But if the value $f(a)$ agrees with our expectations, then we would say that f is continuous at a.

This is a better way to convey the idea of continuity, but it has a problem: It relies on our guessing how f should behave at the point that is omitted from the graph. Perhaps someone from the planet Zog would always guess that if the graph of f looks like this:

around a point a, then the full graph would look like this:

Such a person would end up disagreeing with us most of the time about whether or not a given function is continuous.

On the other hand, someone might have the gift of second sight and when you give them a graph with a point missing, she or he can always guess correctly where the missing point is. She or he would say that every function is continuous.

So this definition leaves room for different opinions on whether a given function is continuous or not. As mathematics is about communication, it is essential that our definition makes it completely clear when a function is continuous, with no ambiguity. So we need to clarify how we arrive at our "expected" value for $f(a)$.

2.2 Rigorous Continuity

One way to decide what $f(a)$ should be uses sequences and limits. If we know the values of a function f for all x near to a point a, but we don't actually know $f(a)$ itself, then we could produce an expected value for $f(a)$ in the following way. If $x_1, x_2, x_3, \ldots$ is a sequence converging to a, then we can take the limit of the sequence $f(x_1), f(x_2), f(x_3), \ldots$ to be the expected value of f at a. This makes things absolutely precise, through the limit concept that we're familiar with.

However, for some functions, different x sequences can lead to different expected values.

Example 2.1

Let $f(x)$ be the function $f(x) = x/|x|$ for $x \neq 0$, i.e.,

$$f(x) = \begin{cases} -1 & \text{if} \quad x < 0, \\ 1 & \text{if} \quad x > 0. \end{cases}$$

Graph of the function $f(x) = x/|x|$

We could use the sequence $x_1, x_2, \ldots$, where $x_i = 1/i$, to find an expected value for $f(0)$, since this sequence converges to 0. The values $f(x_1), f(x_2), \ldots$ are $1, 1, 1, \ldots$. This clearly converges to 1, so we expect the value 1 for $f(0)$.

Alternatively, we could use the sequence where $x_i = -1/i$. This also converges to 0, and $f(x_i)$ is now -1 for all i, so the sequence $f(x_1), f(x_2), \ldots$ converges as well, but to -1. So this sequence leads us to expect $f(0)$ to be -1.

Finally, if we use the sequence $x_i = (-1)^i/i$, so that $x_1, x_2, x_3, \ldots$ is the sequence $-1, 1/2, -1/3, \ldots$, then the f sequence oscillates between -1 and 1 and has no limit. Hence this sequence does not give an expected value for f.

Moreover, for some functions, there may be no x sequence for which the corresponding f sequence converges. In this case we are unable to find an expected value.

In these cases, where different sequences lead to different answers, or where no sequence can be found, we should agree that f is discontinuous, as we cannot find a reasonable expected value. This leaves us with the conclusion that f is continuous at a if, for *every* sequence $x_1, x_2, x_3, \ldots$ which converges to a, the sequence $f(x_1), f(x_2), f(x_3), \ldots$ converges to $f(a)$.

This has removed all ambiguities, is completely rigorous, and agrees with our instincts. But it is rather impractical: We have to consider *all* sequences converging to a and calculate an expected value for $f(a)$ from each one and only if they all agree can we guess what $f(a)$ should be.

Fortunately, we can tidy this up a little using "margins of error" in place of sequences. The idea behind our use of sequences was that as a sequence $x_1, x_2, \ldots$ gets closer and closer to x_0, so the sequence $f(x_1), f(x_2), \ldots$ should get closer and closer to $f(a)$. In other words, as you look at values of x closer and closer to the point a, the values of $f(x)$ should get closer and closer to $f(a)$. In particular, if you specify a certain margin of error, i.e., a small positive number ϵ, then when x is close enough to a, the difference between $f(x)$ and $f(a)$ must be less than the margin of error ϵ. In particular, it should be possible to find some number $\delta > 0$ such that the distance from $f(x)$ to $f(a)$ is less than ϵ whenever the distance from x to a is less than δ.

The process of checking that a function is continuous then becomes a game: If I claim that f is continuous at a then, for any number $\epsilon > 0$ that you choose, I have to be able to find a δ such that $|f(x) - f(a)| < \epsilon$ whenever $|x - a| < \delta$.

Example 2.2

Let f be the function given by $f(x) = x^2$. If I claim that this is continuous at $a = 2$, then for any $\epsilon > 0$ that you choose, I must find a $\delta > 0$ such that $|f(x) - f(a)| < \epsilon$ whenever $|x - a| < \delta$.

Suppose, then, that you choose $\epsilon = 1$. I must find a δ such that $|f(x) - f(2)| < 1$ whenever $|x - 2| < \delta$, i.e., $3 < f(x) < 5$ whenever $2 - \delta < x < 2 + \delta$. So I could choose $\delta = 1/5$ for example, since $2 - \frac{1}{5} < x < 2\frac{1}{5}$ implies that $3\frac{6}{25} < x^2 < 4\frac{21}{25}$.

However, it is important to be able to handle any margin of error ϵ.

Example 2.3

Suppose that f is the function given by

$$f(x) = \begin{cases} x & \text{if } x \neq 1, \\ 1\frac{1}{2} & \text{if } x = 1. \end{cases}$$

With the margin of error $\epsilon = 3/4$, we can find a δ that seems to show that f is continuous at $a = 1$; simply take $\delta = 1/4$. For, if $1 - \delta = 3/4 < x < 1 + \delta = 1\frac{1}{4}$, then $f(x)$ lies between $f(1) - \epsilon$ and $f(1) + \epsilon$, i.e., $f(x)$ lies between $1\frac{1}{2} - \frac{3}{4}$ and $1\frac{1}{2} + \frac{3}{4}$.

However, a tighter margin of error would reveal that this function is not continuous. If you take $\epsilon = 1/4$, then this would require us to find a $\delta > 0$ such that $|f(x) - f(1)| < \epsilon$ whenever $|x - 1| < \delta$, i.e., $|f(x) - 1\frac{1}{2}| < 1/4$ whenever $|x - 1| < \delta$. This is not possible, as the graph shows.

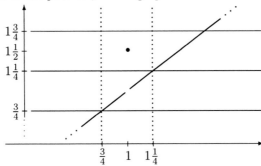

For f to be continuous, then, we need to be able to find a δ for any $\epsilon > 0$. This leads us to the following definition.

Definition: Given any subset S of $\mathbf{R}$, a function $f : S \to \mathbf{R}$ is **continuous at a point** a in its domain S if: For every $\epsilon > 0$, it is possible to find a $\delta > 0$ such that

$$|f(x) - f(a)| < \epsilon \quad \text{whenever} \quad |x - a| < \delta.$$

A function $f : S \to \mathbf{R}$ is said to be **continuous** if it is continuous at every point in S.

This is, at last, a rigorous, practical definition of continuity which matches our intuition.

However, we would like to replace this definition by another approach to continuity. On the one hand, we would like to remove those ϵs and δs which can be difficult to work with. On the other hand, we would like to get away from the notion of distance, which is at the heart of this definition.

2.3 Open Sets

The condition $|x - a| < \delta$ says that x differs from a by less than δ. This can be rephrased as saying that $a - \delta < x < a + \delta$. The set of all values of x satisfying this condition forms a set which we call the "open interval" $(a - \delta, a + \delta)$. In general, if a, b are two real numbers, then the **open interval** (a, b) is the set

$$(a, b) = \{x \in \mathbf{R} : a < x < b\}$$

consisting of all numbers between a and b, excluding a and b themselves. It is convenient to allow a and/or b to be infinite. For example, we may write $(-\infty, 2)$ for the set of all real numbers x such that $x < 2$, and $(3, \infty)$ for the set of all real numbers x such that $x > 3$.

If a and b are both finite, then we can also include the end points to get the **closed interval**

$$[a, b] = \{x \in \mathbf{R} : a \leq x \leq b\}.$$

Mixing these notations, we obtain a **half-open interval**, such as $(2, 3]$, $[2, 3)$, $[2, \infty)$, $(-\infty, 3]$ etc. which include one end point but not the other.

If we take a union of open intervals, then the result will not usually be an interval. For example $(0, 2) \cup (3, 4)$ cannot be written as (a, b) for any pair of numbers a and b. Instead, we call such a union an "open set." Traditionally, this is rephrased in the following way:

Definition: If $S \subset \mathbf{R}$ is a subset of $\mathbf{R}$ then we say S is **open** if, for every point $x \in S$, there is some open interval $(x - \delta_x, x + \delta_x)$ (where $\delta_x > 0$) contained within S.

For example, an open interval such as $(0, 2)$ is open, because any point $x \in (0, 2)$ does have such an open interval around it: If $x \leq 1$, then take $\delta_x = x$, so that $(x - \delta_x, x + \delta_x) = (x - x, x + x) = (0, 2x) \subset (0, 2)$ as $x \leq 1$, and if $x \geq 1$, take $\delta_x = 2 - x$, so that $(x - \delta_x, x + \delta_x) = (x - (2 - x), x + (2 - x)) = (2x - 2, 2) \subset (0, 2)$ as $x \geq 1$. Note that, as in this example, δ_x usually depends on x.

The interval $(x - \delta_x, x + \delta_x)$ about a point x is sometimes called an **open neighbourhood** of x, and we can think of it as "breathing space" for the point x. So an open set is one in which every point has some breathing space.

To see that a union of two open intervals, such as $(0, 2) \cup (3, 4)$, is open by this definition, we can find a δ_x for each $x \in (0, 2) \cup (3, 4)$ in the following way. If $x \in (0, 2)$, we can take $\delta_x = x$ or $\delta_x = 2 - x$ as above, and if $x \in (3, 4)$, then we take $\delta_x = \frac{1}{2}(x - 3)$ or $\delta_x = \frac{1}{2}(4 - x)$ according to whether $x \leq 3\frac{1}{2}$ or $x \geq 3\frac{1}{2}$. And an infinite open interval such as $(2, \infty)$ is open, because we can just take $\delta_x = x - 2$.

On the other hand, a half-open interval, such as $[2, 3)$, is not open, because this contains 2, yet any open interval $(2 - \delta, 2 + \delta)$ will contain a number less than 2 and, hence, cannot be contained in $[2, 3)$.

Similarly, a closed interval such as $[0, 2]$ is not open because if we take $x = 2$, there is no $\delta > 0$ such that $(2 - \delta, 2 + \delta) \subset [0, 2]$.

Since every point x in an open set S belongs to one the open intervals $(x - \delta_x, x + \delta_x)$, and each such interval is contained in S, we can think of S as the union of all these intervals. Hence an open set is indeed a union of open intervals.

Notice that when we think of S as such a union, we are taking the union of an infinite number of open intervals, one for each point in S. This is one example of the fact that open sets behave very well with respect to unions:

Proposition 2.4

The union of any collection of open sets is open.

Proof

Suppose we have a collection $\{S_i\}$ of open sets, indexed by $i \in \mathcal{I}$ for some set $\mathcal{I}$. If x lies in the union $\bigcup_{i \in \mathcal{I}} S_i$, then this says that x lies in one of the sets S_i. Since S_i is open, there is some breathing space around x in S_i, i.e., there is some $\delta_x > 0$ such that $(x - \delta_x, x + \delta_x) \subset S_i$. But if this open interval is contained in S_i, then it is also contained in the union $\bigcup_{i \in \mathcal{I}} S_i$. Hence we have shown that for each x in this union, there is an open interval $(x - \delta_x, x + \delta_x)$ about x contained in the union, i.e., the union is open.   □

On the other hand, we cannot be so relaxed when taking intersections.

Example 2.5

An infinite intersection of open subsets of $\mathbf{R}$ need not be open. Let

$$S_1 = (-1, 1), \quad S_2 = (-1, \frac{1}{2}), \quad S_3 = (-1, \frac{1}{3}), \quad \ldots, \quad S_i = (-1, \frac{1}{i}), \quad \ldots,$$

and let I be their intersection $I = S_1 \cap S_2 \cap S_3 \cap \cdots$. Then I is the interval $(-1, 0]$, which is not open as it has no breathing space around the point 0.

We can, however, take finite intersections safely:

Proposition 2.6

Any finite intersection of open sets is open.

Proof

Let $S_1, \ldots, S_n$ be a finite list of open sets. If x is in the intersection $\bigcap_{i=1}^{n} S_i$, then x belongs to each of the sets S_i. As each is open, we have, for each i, a number $\delta_{x,i} > 0$ such that $(x - \delta_{x,i}, x + \delta_{x,i})$ is contained in S_i. Let $\delta_x = \min(\delta_{x,1}, \ldots, \delta_{x,n})$. Since there are only finitely many numbers $\delta_{x,1}, \ldots, \delta_{x,n}$ and they are all positive, so their minimum will also be positive. And, for each i,

$$(x - \delta_x, x + \delta_x) \subset (x - \delta_{x,i}, x + \delta_{x,i}) \subset S_i.$$

Hence the open interval $(x - \delta_x, x + \delta_x)$ is contained in each S_i and, consequently, is contained in the intersection. Thus we have found some breathing space around x in the intersection, so this intersection is open.   □

Finally, note that there are two extreme examples of open sets. The first is the whole real line $\mathbf{R}$. If $x \in \mathbf{R}$, then we can take δ_x to be any positive number, and $(x - \delta_x, x + \delta_x)$ will be contained in $\mathbf{R}$. So $\mathbf{R}$ itself is an open set.

On the other hand, the empty set $\emptyset$ is also open. This is because the definition of open requires some breathing space for every point in the set. Since there are no points in $\emptyset$, this requirement is automatically satisfied. So $\emptyset$ is an open set in $\mathbf{R}$.

We have discussed open sets at some length, because they form the basis for the topological approach to continuity. However, closed intervals are often useful too, and they generalize to the notion of closed set which we will now look at.

Intuition says that a closed set should be one which is not open, but a half-open interval such as $[0,1)$ is not open, yet it is not closed either. So we need a slightly more complicated definition:

> Definition: A subset $S \subset \mathbf{R}$ is **closed** if its complement $\mathbf{R} - S$ (i.e., the set of all real numbers which are not in S) is open.

Then $[a, b]$ is closed, because its complement is $(-\infty, a) \cup (b, \infty)$ and this is open. (If $x < a$, take $\delta_x = a - x$, and if $x > b$, take $\delta_x = x - b$.)

On the other hand, an open interval such as $(2, 3)$ is not closed, because its complement is $(-\infty, 2] \cup [3, \infty)$. This is not open as there is no breathing space around the point 2. Since the complement is not open, the set $(2, 3)$ is not closed.

Similarly, a finite half-open interval is not closed, although infinite half-open intervals are. For example, $(-\infty, 2]$ is closed, because its complement $(2, \infty)$ is open.

Of all these examples, the finite half-open intervals are the most typical, being neither open nor closed: Most subsets of $\mathbf{R}$ are neither open nor closed.

2.4 Continuity by Open Sets

In order to define continuity in terms of open sets, we will need the following concept.

> Definition: If $f : D \to C$ is a function, and S is a subset of C, then the **preimage** of S under f, written $f^{-1}(S)$, is the subset of D defined by
>
> $$f^{-1}(S) = \{x \in D : f(x) \in S\}.$$

In other words, the preimage of S consists of all points which get mapped to S by f.

Example 2.7

Let $f : \mathbf{R} \to \mathbf{R}$ be the function $f(x) = 4x^2 - 4x$.

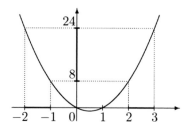

Then:

$f^{-1}(\mathbf{R}) = \mathbf{R}$, $f^{-1}(\emptyset) = \emptyset$,

$f^{-1}[-1, \infty) = \mathbf{R}$, $f^{-1}(-1, \infty) = \mathbf{R} - \{\frac{1}{2}\} = (-\infty, \frac{1}{2}) \cup (\frac{1}{2}, \infty)$,

$f^{-1}[0, \infty) = (-\infty, 0] \cup [1, \infty)$, $f^{-1}(0, \infty) = (-\infty, 0) \cup (1, \infty)$,

$f^{-1}(-\infty, -1] = \{\frac{1}{2}\}$, $f^{-1}(-\infty, -1) = \emptyset$,

$f^{-1}(-\infty, 0] = [0, 1]$, $f^{-1}(-\infty, 0) = (0, 1)$,

$f^{-1}\{0\} = \{0, 1\}$, $f^{-1}\{24\} = \{-2, 3\}$,

$f^{-1}(8, 24) = (-2, -1) \cup (2, 3)$, $f^{-1}[8, 24] = [-2, -1] \cup [2, 3]$.

We see from this example that the preimage of a set can be a single point, or a set of points, or it may even be empty.

WARNING: The notation $f^{-1}(S)$ can be confusing, because it looks like we are assuming that f has an inverse function f^{-1}. This is not the case: We can define preimages for any function, invertible or not.

Preimages behave very nicely with respect to other set operations, as the following result shows.

Lemma 2.8

If $f : S \to T$ and U, V are subsets of T, then

i. $f^{-1}(U \cap V) = f^{-1}(U) \cap f^{-1}(V)$ ii. $f^{-1}(U \cup V) = f^{-1}(U) \cup f^{-1}(V)$

iii. $f^{-1}(\emptyset) = \emptyset$ iv. $f^{-1}(T) = S$

Proof

If $x \in f^{-1}(U \cap V)$, then $f(x) \in U \cap V$, so $f(x) \in U$ and $f(x) \in V$. Hence $x \in f^{-1}(U)$ and $x \in f^{-1}(V)$, i.e., $x \in f^{-1}(U) \cap f^{-1}(V)$. Conversely, if $x \in$

$f^{-1}(U) \cap f^{-1}(V)$, then $x \in f^{-1}(U \cap V)$ by a similar argument. Part (ii) is proved similarly.

The preimage $f^{-1}(\emptyset)$ is always empty, because if it contained any element s, then $f(s) \in \emptyset$, which cannot happen. And $f^{-1}(T) = S$, because $f(s) \in T$ for every element $s \in S$. $\qquad\square$

Now we can state the open-set approach to continuity.

Theorem 2.9

Let $f : \mathbf{R} \to \mathbf{R}$ be a function. If f is continuous, then $f^{-1}(S)$ is open whenever $S \subset \mathbf{R}$ is open. And vice versa: If $f^{-1}(S)$ is open whenever S is an open subset of $\mathbf{R}$, then f is continuous.

Proof

For the first part we assume that f is continuous, so for every point $a \in \mathbf{R}$, and every $\epsilon > 0$, there is some $\delta > 0$ such that

$$x \in (a - \delta, a + \delta) \implies f(x) \in (f(a) - \epsilon, f(a) + \epsilon).$$

Now let S be an open subset of $\mathbf{R}$. To show that $f^{-1}(S)$ is open, let a be any point in $f^{-1}(S)$, and we will find a number δ such that $(a - \delta, a + \delta) \subset f^{-1}(S)$. If $a \in f^{-1}(S)$, then $f(a) \in S$ and so, as S is open, there is some $\epsilon > 0$ such that $(f(a) - \epsilon, f(a) + \epsilon) \subset S$. Since f is continuous, this means that we can find a δ such that $f(x) \in (f(a) - \epsilon, f(a) + \epsilon)$ whenever $x \in (a - \delta, a + \delta)$. Since $(f(a) - \epsilon, f(a) + \epsilon) \subset S$, we see that $f(x) \in S$ whenever $x \in (a - \delta, a + \delta)$. That means $(a - \delta, a + \delta) \in f^{-1}(S)$, which is just what we needed to show that $f^{-1}(S)$ is open.

Now, to prove the converse, suppose that $f^{-1}(S)$ is open whenever S is open. Let a be any point in the domain of f and ϵ any positive real number. The interval $(f(a) - \epsilon, f(a) + \epsilon)$ is open, so its preimage under f is also open. This preimage contains a, because $f(a)$ is contained in $(f(a) - \epsilon, f(a) + \epsilon)$. So, as it is open, we can find an interval $(a - \delta, a + \delta)$ contained in the preimage. Thus

$$x \in (a - \delta, a + \delta) \subset f^{-1}(f(a) - \epsilon, f(a) + \epsilon) \implies f(x) \in (f(a) - \epsilon, f(a) + \epsilon),$$

i.e., f is continuous. $\qquad\square$

Example 2.10

Let $f : \mathbf{R} \to \mathbf{R}$ be the function $f(x) = 2x + 3$. To see that f is continuous using this open-set approach, let $U \subset \mathbf{R}$ be any open set; we must show that $f^{-1}(U)$ is open. So take any point $x \in f^{-1}(U)$. Then $2x + 3 = f(x) \in U$ and, as U is open, we can find an $\epsilon > 0$ such that $(2x + 3 - \epsilon, 2x + 3 + \epsilon) \subset U$. Now

$$(2x + 3 - \epsilon, 2x + 3 + \epsilon) = (2(x - \frac{\epsilon}{2}) + 3, 2(x + \frac{\epsilon}{2}) + 3).$$

So if $x' \in (x - \frac{\epsilon}{2}, x + \frac{\epsilon}{2})$, then $f(x') \in (2x + 3 - \epsilon, 2x + 3 + \epsilon) \subset U$. In other words, the interval $(x - \frac{\epsilon}{2}, x + \frac{\epsilon}{2})$ is contained in $f^{-1}(U)$. Thus x has some breathing space around it, i.e., $f^{-1}(U)$ is open and, hence, f is continuous.

EXERCISES

2.1. Prove that the set $(2, 3)$ is open, by giving, for each point $x \in (2, 3)$, a formula for $\delta_x > 0$ such that $(x - \delta_x, x + \delta_x)$ is contained in $(2, 3)$.

2.2. Prove that the interval $[-1, 1]$ is closed, by proving that its complement in $\mathbf{R}$ is open.

2.3. Show that any subset $\{x\}$, containing a single number x, is closed.

2.4. Prove that the subset $\mathbf{Z} \subset \mathbf{R}$, consisting of all the integers, is closed.

2.5. Let $f : \mathbf{R} \to \mathbf{R}$ be the function given by $f(x) = x^3 - 3x$. Calculate the preimages $f^{-1}[-2, 2]$, $f^{-1}(2, 18)$, $f^{-1}[2, 18)$, $f^{-1}[0, 2]$.

2.6. If $g(x) = \cos(x)$, what are $g^{-1}(1)$, $g^{-1}[0, 1]$, $g^{-1}[-1, 1]$?

2.7. Let $f : \mathbf{R} \to \mathbf{R}$ be the function $f(x) = 3 - 2x$.

   – Prove that f is continuous using the ϵ, δ definition of continuity.

   – Prove that f is continuous by showing that the preimage $f^{-1}(S)$ of every open set $S \subset \mathbf{R}$ is open.

3
Topological Spaces

So far, we have only considered functions on the real line. We have seen how to hide those annoying ϵs and δs in the definition of continuity, replacing them with open sets. This enables us to consider functions with domains and ranges different from **R**; all we need is some notion of "open set".

3.1 Topological Spaces

Definition: A **topological space** is a set, X, together with a collection, $\mathcal{T}$, of subsets of X, called "open" sets, which satisfy the following rules:

T1. The set X itself is "open"

T2. The empty set is "open"

T3. Arbitrary unions of "open" sets are "open"

T4. Finite intersections of "open" sets are "open"

The collection of "open" sets is called the **topology** on X. The conditions T1 – T4 are modelled on the properties of open sets in **R** that we observed in the previous chapter. They ensure that the "open" sets of any topological space behave somewhat like the open subsets of **R**. So, of course, our first example of a topological space is **R** itself.

M.D. Crossley, *Essential Topology*, Springer Undergraduate
Mathematics Series, DOI 10.1007/978-1-84628-194-5_3,
© Springer-Verlag London Limited 2010

Example 3.1

The real line, **R**, with the open sets defined in Section 2.3, is a topological space, as Propositions 2.4 and 2.6, and the comments following Proposition 2.6, show.

Example 3.2

If B is the set $\{0,1\}$ consisting of just two elements, then we can make this a topological space in a few different ways.

Firstly, we could agree that only the empty set $\emptyset$ and the whole set $\{0,1\}$ are to be called open. This satisfies axioms T1 and T2. T3 is also satisfied because the only possible union of open sets is where we take $\emptyset \cup \{0,1\}$ and the result here is $\{0,1\}$ which, we have agreed, is open. Finally, T4 is also true, for the only intersection is $\emptyset \cap \{0,1\}$ which is $\emptyset$ which, we've agreed, is open.

Example 3.3

On the other hand, we can topologize $\{0,1\}$ by calling all the subsets $\emptyset$, $\{0\}$, $\{1\}$, $\{0,1\}$ open. The axioms are then satisfied, because the empty set and the whole set are included in our list of open sets, and any intersections or unions will be open since all subsets are open.

Example 3.4

In fact, given any set S, there are at least two ways of making S a topological space, illustrated by the preceding examples. On the one hand, we can call only the empty set and S itself open. This is called the **indiscrete topology** on the set S.

And we can take all the subsets of S to be open. This is called the **discrete topology**.

The topology on a set gives a meaning to the phrase "open subset". Given that, we can define the phrase "closed subset" in terms of this, much as we did for **R**. We say that a subset S of a topological space T is **closed** if the complement $T - S$, consisting of all elements of T which are not in S, is open.

Example 3.5

In the set $\{0,1\}$ with the topology of Example 3.2, where only $\emptyset$ and $\{0,1\}$ are open, the only closed sets are $\{0,1\}$ and $\emptyset$. The other subsets, $\{0\}$ and $\{1\}$, are not closed, as their complements are not open.

Example 3.6

If we give $\{0,1\}$ the topology of Example 3.3, where every subset is open, then every subset will also be closed.

  Note that, in these examples, there are subsets which are both open and closed, such as $\emptyset$ and $\{0,1\}$, and there are subsets which are neither open nor closed, such as $\{0\}$ in the first example. So, just as with $\mathbf{R}$, the terms "open" and "closed" are not opposite to each other, and we can only know if a subset is closed or not by looking at its complement.

  We define the term "topological space" so as to be able to give the following more general definition of continuity.

Definition: A function $f : S \to T$ from one topological space to another is **continuous** if the preimage $f^{-1}(Q)$ of every open set $Q \subset T$ is an open set in S.

It is common to use the shorter word **map** instead of the word "function", and, when speaking of a map between two topological spaces, the term "map" usually means a continuous function.

Example 3.7

Let $B = \{0,1\}$ have the discrete topology, and define $f : B \to \mathbf{R}$ by $f(0) = -1$, $f(1) = 1$. To check whether or not f is continuous, let U be any open set in $\mathbf{R}$. Then the preimage of U is given by

$$f^{-1}(U) = \begin{cases} \{0\} & \text{if } -1 \in U \text{ and } 1 \notin U, \\ \{1\} & \text{if } -1 \notin U \text{ and } 1 \in U, \\ \{0,1\} & \text{if } -1 \in U \text{ and } 1 \in U, \\ \emptyset & \text{if } -1 \notin U \text{ and } 1 \notin U. \end{cases}$$

In each case, the preimage of U is open, since every subset of B is open in the discrete topology. Hence f is continuous.

Example 3.8

Let $B = \{0,1\}$ with the discrete topology and let $T = \{0,1\}$ with the indiscrete topology. Define $g : B \to T$ by $g(0) = 0$, $g(1) = 1$. It is easy to check that g is continuous, since the only open sets in T are $\emptyset$ and T, and $g^{-1}(\emptyset) = \emptyset$ while $g^{-1}(T) = B$, both of which are open.

  However, if $h : T \to B$ is the function $h(0) = 0$ and $h(1) = 1$, then h is not continuous. For $\{0\}$ is an open set in B, but $h^{-1}\{0\} = \{0\}$ is not an open set in T.

As these examples show, since continuity is defined in terms of open sets, if we change the topology then we change the notion of continuity. Examples 3.7 and 3.8 are also instances of a more general phenomenon described by the following proposition.

Proposition 3.9

If S has the discrete topology and T is any topological space, then any function $f : S \to T$ is continuous.

  If T has the indiscrete topology and S is any topological space, then any function $f : S \to T$ is continuous.

Proof

If $f : S \to T$ is to be continuous, the preimage of any open set must be open. But if S has the discrete topology, then every subset of S is open, so in particular, every preimage of an open set must be open. Thus f is continuous.

  If $f : S \to T$ is to be continuous where T has the indiscrete topology, then the preimage of any open set in T must be open in S. But the only open sets in T are the empty set and the whole set T. By Lemma 2.8, the preimage of $\emptyset \subset T$ is $\emptyset \subset S$, which is open. And the preimage of $T \subset T$ is the whole of S, which is also open.   □

  Before we move on, we remark that composition respects continuity:

Proposition 3.10

If R, S, T are topological spaces and $f : R \to S$, $g : S \to T$ are continuous functions, then $g \circ f : R \to T$ is continuous.

Proof

Let $U \subset T$ be an open set. As g is continuous, $g^{-1}(U)$ is open and hence, as f is continuous, $f^{-1}(g^{-1}(U))$ is an open set in R. Now $(g \circ f)^{-1}(U) = f^{-1}(g^{-1}(U))$ since

$$(g \circ f)^{-1}(U) = \{r \in R : g \circ f(r) \in U\} = \{r \in R : g(f(r)) \in U\}$$
$$= \{r \in R : f(r) \in g^{-1}(U)\} = f^{-1}(g^{-1}(U)).$$

Hence $(g \circ f)^{-1}(U)$ is open whenever U is, i.e., $g \circ f$ is continuous.   □

3.2 More Examples of Topological Spaces

In order to construct more interesting examples of topological spaces, we need to be able to use higher dimensions. We can topologize $\mathbf{R}^2$ as follows.

Example 3.11

We put a topology on $\mathbf{R}^2$ in a similar way to $\mathbf{R}$. For any point (x, y) in $\mathbf{R}^2$ and real number $\delta > 0$, let

$$B_\delta(x, y) = \{(x', y') \in \mathbf{R}^2 : \sqrt{(x' - x)^2 + (y' - y)^2} < \delta\}.$$

We call this the **open ball** of radius δ around (x, y); it is analogous to the open interval $(x - \delta, x + \delta)$ in $\mathbf{R}$. A subset of $Q \subset \mathbf{R}^2$ is defined to be open if, for every $(x, y) \in Q$, there is some $\delta > 0$ such that $B_\delta(x, y)$ is contained in Q.

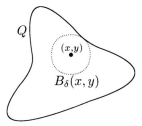

The proof that this topology satisfies the axioms is much the same as for $\mathbf{R}$ and is left as Exercise 3.5.

To give some evidence that this is the "right" topology, we will look at some familiar maps from $\mathbf{R}^2$ to $\mathbf{R}$.

Example 3.12

The addition map $A : \mathbf{R}^2 \to \mathbf{R}$, $A(x, y) = x + y$ is continuous. To see this, let U be an open set in $\mathbf{R}$, so we must show that $A^{-1}(U)$ is open. Let (a, b) be any point in $A^{-1}(U)$, so $a + b = A(a, b) \in U$. As U is open, there is some δ such that the interval $(a + b - \delta, a + b + \delta)$ is contained in U and, hence, $A^{-1}(a + b - \delta, a + b + \delta) \subset A^{-1}(U)$. The set $A^{-1}(a + b - \delta, a + b + \delta)$ is the region of $\mathbf{R}^2$ between (but not including) the two lines $y = a + b - \delta - x$ and $y = a + b + \delta - x$.

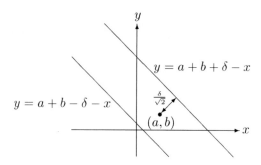

Now (a, b) is equally spaced between the two boundary lines, so the distance from (a, b) to the nearest point on the boundary is $d = \delta/\sqrt{2}$. Hence $B_d(a, b) \subset A^{-1}(a+b-\delta, a+b+\delta) \subset A^{-1}(U)$. So around the point $(a, b) \in A^{-1}(U)$ there is an open ball contained in $A^{-1}(U)$ and hence $A^{-1}(U)$ is open in $\mathbf{R}^2$, as required. Thus A is continuous.

Example 3.13

Similarly, multiplication $\mathbf{R}^2 \to \mathbf{R}$ is continuous.

Example 3.14

The diagonal map $d : \mathbf{R} \to \mathbf{R}^2$ defined by $d(x) = (x, x)$ is continuous. To see this, let $Q \subset \mathbf{R}^2$ be open and $x \in d^{-1}(Q)$. Thus $d(x) = (x, x) \in Q$, and there is some $\epsilon > 0$ such that $B_\epsilon(x, x)$ is contained in Q. Let $\delta = \epsilon/\sqrt{2}$, and let y be any point in the interval $(x - \delta, x + \delta) \subset \mathbf{R}$. Then the distance from $d(y) = (y, y)$ to $d(x) = (x, x)$ is $\sqrt{(x - y)^2 + (x - y)^2} < \sqrt{\delta^2 + \delta^2} = \epsilon$, so $d(y) \in B_\epsilon(x, x)$.

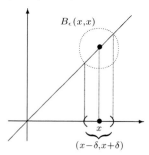

Hence $d(y) \in Q$. Consequently, $(x - \delta, x + \delta) \subset d^{-1}(Q)$, so $d^{-1}(Q)$ is open. Thus the preimage of any open subset of $\mathbf{R}^2$ is open in $\mathbf{R}$, i.e., d is continuous.

Corollary 3.15

Any polynomial $f : \mathbf{R} \to \mathbf{R}$, $f(x) = a_0 + a_1 x + \cdots + a_n x^n$ is continuous.

Proof

The function f can be expressed as a composite of multiplications, additions and inclusions which, by the preceding examples and Proposition 3.10, is continuous.   □

Example 3.16

We can topologize n-dimensional Euclidean space, $\mathbf{R}^n$, in the same way as $\mathbf{R}^2$, using the open ball

$$B_\delta(\mathbf{x}) = \{\mathbf{y} \in \mathbf{R}^n : d(\mathbf{x}, \mathbf{y}) < \delta\},$$

where $d(\mathbf{x}, \mathbf{y}) = \sqrt{(x_1 - y_1)^2 + \cdots + (x_n - y_n)^2}$ if $\mathbf{x} = (x_1, \ldots, x_n)$ and $\mathbf{y} = (y_1, \ldots, y_n)$. We then say that a subset of $\mathbf{R}^n$ is open if, around every point in the subset, we can find an open ball contained in the subset.

Notice how we have put a topology on $\mathbf{R}^n$ using only the concept of distance between two points. We can do the same in any set which has a reasonable notion of distance; such sets are called **metric spaces**. This is one way of generating examples of topological spaces. Another is by topologizing subsets of known topological spaces.

If T is any topological space (for example, $\mathbf{R}^n$), and $S \subset T$ is any subset of T, then the **subspace topology**[1] on S is defined as follows: A subset of S is said to be open (in the subspace topology) if it is the intersection of S with an open set in T. With these open sets, S becomes a topological space in its own right, and we refer to S as a **subspace** of T.

For example, in the diagram below, U is an open subset of the space T. The shaded area, which is $S \cap U$, is an open set in the subspace topology on the subset S.

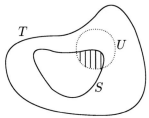

Example 3.17

The set of integers forms a subset of $\mathbf{R}$, usually denoted by $\mathbf{Z}$, so we can give it the subspace topology. It turns out that every subset of $\mathbf{Z}$ is open. For if $n \in \mathbf{Z}$,

---

[1] The subspace topology is also called the "weak" topology or "induced" topology.

then $\{n\} = \mathbf{Z} \cap (n - \frac{1}{2}, n + \frac{1}{2})$, the intersection of $\mathbf{Z}$ with an open set of $\mathbf{R}$. So $\{n\}$ is open in $\mathbf{Z}$ and, since any union of open sets is open, so any subset of $\mathbf{Z}$ is open. In other words, the subspace topology on $\mathbf{Z}$ is the same as the discrete topology. So we say that $\mathbf{Z}$ is a **discrete space**. Proposition 3.9 then tells us that every function whose domain is $\mathbf{Z}$ is continuous.

Example 3.18

The set of rationals is another subset of $\mathbf{R}$, denoted by $\mathbf{Q}$. With the subspace topology this is a very interesting space. It looks like it might be discrete, because between every pair of rationals there is an irrational number. However, it is not discrete, since any subset containing just a single rational is not open. For example, if $n \in \mathbf{Q}$, then any open subset of $\mathbf{Q}$ containing n is of the form $\mathbf{Q} \cap S$, where S is an open subset of $\mathbf{R}$ containing n. Since S is open, it contains some interval $(n - \delta_n, n + \delta_n)$ around n, where $\delta_n > 0$. Any such interval will contain some rationals other than n, so we cannot express $\{n\}$ as $\mathbf{Q} \cap S$ for any open subset $S \subset \mathbf{R}$.

Example 3.19

Let S^1 be the subset of $\mathbf{R}^2$ consisting of all points on the circle of radius 1 around the origin, i.e.,

$$S^1 = \{(x, y) \in \mathbf{R}^2 : x^2 + y^2 = 1\},$$

with the subspace topology. This says that a subset of S^1 is open if it is the intersection of S^1 with an open set in $\mathbf{R}^2$. So, for example, one open set in $\mathbf{R}^2$ is the open ball $B_1(1, 1)$ and hence one open set in S^1 will be the intersection $S^1 \cap B_1(1, 1)$ which is the quarter-circle between 12 o'clock and 3 o'clock, excluding the end points (12 and 3 o'clock), as depicted below.

Example 3.20

The 2-sphere S^2 is the analogous subset of $\mathbf{R}^3$,

$$S^2 = \{(x, y, z) \in \mathbf{R}^3 : x^2 + y^2 + z^2 = 1\},$$

with the subspace topology, and looks like:

Example 3.21

Similarly, we can take the subset consisting of all points with distance 1 from the origin in $\mathbf{R}^{n+1}$, with the subspace topology. We call this space the n-**sphere**, denoted by S^n.

We can even allow $n = 0$: S^0 is the set of all points 1 away from the origin 0 in $\mathbf{R}$, so S^0 consists of just two points $\{+1, -1\}$.

The reason we write S^n, rather than S^{n+1}, for the sphere in $\mathbf{R}^{n+1}$ is because, locally, S^n looks like $\mathbf{R}^n$. For example, near the North Pole, the circle S^1 looks just like a line, and S^2 looks just like a plane.

Example 3.22

Another useful subset of $\mathbf{R}^2$ is the set of non-zero points, $\mathbf{R}^2 - \{0\}$, sometimes written $\mathbf{C}^\times$, as it corresponds to the set of invertible complex numbers. We can think of $\mathbf{R}^2 - \{0\}$ as a topological space by using the subspace topology.

Similarly, we can take $\mathbf{R}^n - \{0\}$ for any n, and topologize this as a subspace of $\mathbf{R}^n$. Since $\mathbf{R}^n - \{0\}$ is, itself, an open subset of $\mathbf{R}^n$, any open set of $\mathbf{R}^n - \{0\}$ in the subspace topology is an intersection $\mathbf{R}^n - \{0\} \cap S$ of two open subsets of $\mathbf{R}^n$. Hence every open set of $\mathbf{R}^n - \{0\}$ is an open set of $\mathbf{R}^n$. Conversely, every open set of $\mathbf{R}^n$ which is contained in $\mathbf{R}^n - \{0\}$ is an open set of $\mathbf{R}^n - \{0\}$.

Example 3.23

The space $\mathbf{R} - \{0\}$ is the natural domain of the reciprocal function $x \mapsto 1/x$ and, considering this as a function $f : \mathbf{R} - \{0\} \to \mathbf{R}$, it can be seen to be continuous. For if Q is an open subset of $\mathbf{R}$, then we can show that $f^{-1}(Q)$ is open as follows. Suppose that $x \in f^{-1}(Q)$, i.e., $1/x \in Q$. Since Q is open, there is some $\epsilon > 0$ such that $(\frac{1}{x} - \epsilon, \frac{1}{x} + \epsilon) \subset Q$. We may assume that $\epsilon < |1/x|$, so that $(\frac{1}{x} - \epsilon, \frac{1}{x} + \epsilon)$ does not contain 0. Then

$$f^{-1}\left(\frac{1}{x} - \epsilon, \frac{1}{x} + \epsilon\right) = \left(\frac{1}{\frac{1}{x} + \epsilon}, \frac{1}{\frac{1}{x} - \epsilon}\right) = \left(x - \frac{x^2\epsilon}{1 + x\epsilon}, x + \frac{x^2\epsilon}{1 - x\epsilon}\right).$$

Let $\delta = \min\left(\frac{x^2\epsilon}{1+x\epsilon}, \frac{x^2\epsilon}{1-x\epsilon}\right)$, so that

$$(x - \delta, x + \delta) \subset f^{-1}(\frac{1}{x} - \epsilon, \frac{1}{x} + \epsilon) \subset f^{-1}(Q).$$

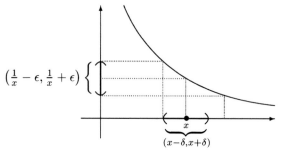

Hence, around every point in $f^{-1}(Q)$ there is some breathing space, so this preimage is open.

Example 3.24

Let $T \subset \mathbf{R}^2$ be the set of points (x, y) with $y \neq 0$, with the subspace topology. Division defines a function $d : T \to \mathbf{R}$ and, by Examples 3.23 and 3.13, we see that d is continuous.

The subspace topology sometimes gives rise to unexpected open sets.

Example 3.25

Let $I = [0, 1]$ be the closed interval in $\mathbf{R}$. With the subspace topology we can view $[0, 1]$ as a topological space in its own right. Then any interval (a, b) with $0 \leq a < b \leq 1$ is open, but also an interval $[0, b)$ with $b \leq 1$ is open as a subset of $[0, 1]$, since $[0, b)$ is the intersection $[0, 1] \cap (-1, b)$ of $[0, 1]$ with the open interval $(-1, b) \subset \mathbf{R}$. Similarly, any interval $(a, 1] = [0, 1] \cap (a, 2)$ is open. And, of course, $[0, 1]$ is open, too.

So we need to be clear which topological space we are talking about when we say that a certain set is open.

Example 3.26

If we take a ring-shaped subset of $\mathbf{R}^3$, and define T^2 to be the set of all points on its boundary, with the subspace topology then we get a **torus**.

To be precise, we can describe the torus T^2 as the set of points (x, y, z) in $\mathbf{R}^3$ which satisfy

$$x^2 + y^2 + z^2 - 4\sqrt{(x^2 + y^2)} = -3.$$

Example 3.27

Now suppose we take a torus, but slice off one side. Put another, matching, sliced torus next to it, so that the two sliced holes face each other. Then if we bring these two sliced tori together until the holes touch, we will get a tubular figure of eight. The resulting topological space, G_2, is called a **surface of genus two**, which is a fancy way of saying that it has two holes.

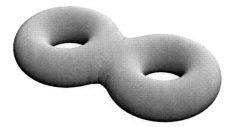

Example 3.28

We can form a **cylinder** by taking all points (x, y, z) in $\mathbf{R}^3$ satisfying

$$x^2 + y^2 = 1 \quad \text{and} \quad 0 \le z \le 1.$$

Example 3.29

Now if we cut the cylinder along the line $x = -1$, $y = 0$, and give the resulting ribbon a half twist, then we can stick the ends back together and get a **Möbius band**. This is famous for only having one side.

Cutting and pasting a subset of $\mathbf{R}^3$ in this way does not necessarily give another subset of $\mathbf{R}^3$. However, in this case it does, and we can topologize the Möbius band as the following subset of $\mathbf{R}^3$:

$$M = \{(-(3+t\sin\theta)\cos(2\theta), (3+t\sin\theta)\sin(2\theta), t\cos\theta) : 0 \le \theta \le \pi, -1 \le t \le 1\}.$$

Example 3.30

The set of all invertible 3×3 real matrices forms a set called $\mathrm{GL}(3, \mathbf{R})$, the **general linear group**. Using the 9 entries in such a matrix as coordinates, we can think of $\mathrm{GL}(3, \mathbf{R})$ as a subset of $\mathbf{R}^9$, and use the subspace topology to make $\mathrm{GL}(3, \mathbf{R})$ into a topological space. Similarly, the set $\mathrm{GL}(n, \mathbf{R})$ of invertible $n \times n$ real matrices can be topologized as a subspace of $\mathbf{R}^{n^2}$.

Example 3.31

Inside $\mathrm{GL}(3, \mathbf{R})$ there is the subset $\mathrm{O}(3)$ of orthogonal matrices, i.e., those matrices P satisfying $P^T P = I$. This subset can again be topologized using the subspace topolgy. The set $\mathrm{O}(3)$ is known as the **orthogonal group**. This is the set of all angle-preserving linear transformations $\mathbf{R}^3 \to \mathbf{R}^3$, such as rotations and reflections. In fact, with a little effort you can check that every matrix in $\mathrm{O}(3)$ is either a rotation, or a reflection, or a combination of both.

Example 3.32

Pursuing this line of thought further, if P is an orthogonal matrix, then

$$\det(P)^2 = \det(P^T)\det(P) = \det(P^T P) = \det(I) = 1$$

so $\det(P)$ is either $+1$ or -1. The subset of 3×3 orthogonal matrices of determinant $+1$ is important and is called $\mathrm{SO}(3)$, the **special orthogonal group**. This is the set of all orientation-preserving, angle-preserving linear transformations of $\mathbf{R}^3$. Once you have verified that $\mathrm{O}(3)$ consists of rotations, reflections and combinations of these, it is easy to see that all matrices in SO(3) are rotations, since reflections have determinant -1.

Example 3.33

A very interesting space arises by considering the set of all straight lines through the origin in $\mathbf{R}^3$, which we write $\mathbf{RP}^2$ for. This is called the **real projective plane**. It is not a subset of $\mathbf{R}^3$, since the elements of $\mathbf{RP}^2$ are not points in $\mathbf{R}^3$ but subsets of $\mathbf{R}^3$. So we cannot use the subspace topology to make it into a topological space.

Instead, we could topologize it as follows. If we take a subset of $\mathbf{RP}^2$, i.e., a collection of lines in $\mathbf{R}^3$, then we can take the union of these lines to get a subset of $\mathbf{R}^3$. We could then define the subset of $\mathbf{RP}^2$ to be open if the corresponding subset of $\mathbf{R}^3$ is open, i.e., $S \subset \mathbf{RP}^2$ is open if $\bigcup_{l \in S} l \subset \mathbf{R}^3$ is open. Unfortunately, this gives the indiscrete topology on $\mathbf{RP}^2$, since the union will contain 0, unless S is empty, but will not contain an open ball around 0 unless $S = \mathbf{RP}^2$.

To avoid this problem, we omit 0, and we say that a subset $S \subset \mathbf{RP}^2$ is open if the subset $\bigcup_{l \in S} (l - 0)$ of $\mathbf{R}^3 - \{0\}$ is open. The empty subset of $\mathbf{RP}^2$ is then open, because the corresponding subset of $\mathbf{R}^3 - \{0\}$ is also the empty set. And the whole set $\mathbf{RP}^2$ is open, because this corresponds to the whole set $\mathbf{R}^3 - \{0\}$.

Unions and intersections of subsets of $\mathbf{RP}^2$ correspond to unions and intersections of subsets of $\mathbf{R}^3 - \{0\}$ so, because the open sets in $\mathbf{R}^3 - \{0\}$ form a topology, these open sets in $\mathbf{RP}^2$ also form a topology.

Example 3.34

We can do something similar in $\mathbf{R}^n$ for any n: Define $\mathbf{RP}^{n-1}$ to be the set of lines through the origin in $\mathbf{R}^n$, with the topology defined in the same way as for $\mathbf{RP}^2$. These spaces $\mathbf{RP}^m$ are called **real projective spaces**.

The space $\mathbf{RP}^0$ is the set of lines through the origin in $\mathbf{R}$. But there is just one such line, so $\mathbf{RP}^0$ is a single point.

Later on, in Chapter 5, we will see that $\mathbf{RP}^1$ is topologically the "same" as the circle S^1. This gives you some idea why we write $\mathbf{RP}^1$ and not $\mathbf{RP}^2$.

A much more startling fact is that $\mathbf{RP}^3$ is the "same" as the space SO(3). We will prove this fact in Proposition 5.67.

3.3 Continuity in the Subspace Topology

As we have seen, the topology determines which maps are continuous. So if the subspace topology is to be useful, we had better check that it gives a sensible notion of continuity. This is expressed by the following result.

Proposition 3.35

Let S and T be topological spaces and $f : S \to T$ a continuous map. Suppose that $Q \subset S$ is a subset whose image, under f, is contained in a subset $R \subset T$, so that f can be restricted to a function $f|_Q : Q \to R$. If Q and R are given the subspace topologies, then $f|_Q$ is continuous.

$$
\begin{array}{ccc}
S & \xrightarrow{\ \ f\ \ } & T \\
\cup & & \cup \\
Q & \xrightarrow{\ f|_Q\ } & R
\end{array}
$$

Proof

Let $P \subset R$ be an open set in the subspace topology. Then $f|_Q^{-1}(P) = Q \cap f^{-1}(P)$. As P is open in the subspace topology, there must be an open set $U \subset T$ such that $P = R \cap U$, in which case

$$f^{-1}(P) = f^{-1}(R \cap U) = f^{-1}(R) \cap f^{-1}(U),$$

so

$$f|_Q^{-1}(P) = Q \cap f^{-1}(R) \cap f^{-1}(U) = Q \cap f^{-1}(U)$$

since $Q \subset f^{-1}(R)$ (by our assumption that the image of Q under f is contained in R). Now, as f is continuous, and U is open, $f^{-1}(U)$ must also be open. So $f|_Q^{-1}(P) = Q \cap f^{-1}(U)$ is an open set in the subspace topology on Q. Hence the preimage under $f|_Q$ of any open set in R is open, i.e., $f|_Q$ is continuous. $\square$

In other words, if a function on, say, S^2 is just a restriction of a function defined on $\mathbf{R}^3$, then the function on S^2 will be continuous if the function on $\mathbf{R}^3$ is continuous.

Example 3.36

The inclusion map $S^2 \hookrightarrow \mathbf{R}^3$ is continuous, as it is simply a restriction of the identity map $(x, y, z) \mapsto (x, y, z)$.

Example 3.37

The determinant function $\mathrm{GL}(3, \mathbf{R}) \to \mathbf{R}$ is continuous. For we have topologized $\mathrm{GL}(3, \mathbf{R})$ as a subspace of $\mathbf{R}^9$, and the determinant function is the restriction of the function $\mathbf{R}^9 \to \mathbf{R}$ given by

$$(a, b, c, d, e, f, g, h, i) \mapsto aei - afh - bdi + bfg + cdh - ceg.$$

This is continuous, being a composite of additions and multiplications, both of which are continuous operations, by Examples 3.12 and 3.13. Hence the restriction to $\mathrm{GL}(3, \mathbf{R})$ is also continuous.

Example 3.38

The function $\mathrm{inv} : \mathrm{GL}(n, \mathbf{R}) \to \mathrm{GL}(n, \mathbf{R})$ that sends a matrix M to its inverse M^{-1} is continuous. For example, if $n = 2$, inv is given by

$$\mathrm{inv} \begin{bmatrix} a & b \\ c & d \end{bmatrix} = \frac{1}{(ad - bc)} \begin{bmatrix} d & -b \\ -c & a \end{bmatrix}.$$

The function

$$\mathrm{adj} \begin{bmatrix} a & b \\ c & d \end{bmatrix} = \begin{bmatrix} d & -b \\ -c & a \end{bmatrix}$$

is the restriction of a continuous map $\mathbf{R}^4 \to \mathbf{R}^4$, hence is continuous by Proposition 3.35. Then inv is given by $\mathrm{inv}(M) = \mathrm{adj}(M)/\det(M)$. Now $\det(M)$ is non-zero if $M \in \mathrm{GL}(n, \mathbf{R})$, and adj and det are continuous and Example 3.24 tells us that division by non-zero numbers is continuous. Hence inv is a composite of continuous functions and, hence, continuous.

For $n > 2$, the same argument applies; the functions adj and det are more complicated, but they are polynomials in the matrix entries, so they are continuous. Hence inv is also continuous.

Example 3.39

Matrix multiplication takes a pair of matrices and returns another matrix. Restricting to 3×3 matrices, we can think of a pair of matrices as being an element of $\mathbf{R}^{18}$, since there are 18 entries between the two matrices, so multiplication corresponds to a function $\mathbf{R}^{18} \to \mathbf{R}^9$. Since this can be expressed in terms of multiplications and additions of real numbers, it is a continuous function. Hence the restriction to multiplication of matrices in $\mathrm{GL}(3, \mathbf{R})$ is also continuous.

These two examples tell us that $GL(3, \mathbf{R})$ forms a **topological group**, i.e., a group where all of the structure maps are continuous. The same is true for $O(3)$ and $SO(3)$.

Example 3.40

We can define a map $r : S^2 \to SO(3)$ by

$$
r(x, y, z) = \begin{bmatrix} 2x^2 - 1 & 2xy & 2xz \\ 2xy & 2y^2 - 1 & 2yz \\ 2xz & 2yz & 2z^2 - 1 \end{bmatrix}.
$$

This is the restriction of a function $\mathbf{R}^3 \to \mathbf{R}^9$; it is straightforward (but tedious) to show that $r(x, y, z)$ is an orthogonal matrix of determinant 1 if $x^2 + y^2 + z^2 = 1$, so that the image of r is indeed contained in $SO(3)$.

The original function $\mathbf{R}^3 \to \mathbf{R}^9$ is continuous because it is simply a composite of additions and multiplications and so, by Proposition 3.35, its restriction, r, is continuous.

We can describe r geometrically as follows. A point (x, y, z) in S^2 determines an axis of rotation, and $r(x, y, z)$ is the matrix which rotates points about this axis by $180°$. To check this, try writing down the matrix describing rotation by $180°$ about the line through (x, y, z). If $x^2 + y^2 + z^2 = 1$ (i.e., $(x, y, z) \in S^2$), then it will be exactly the matrix given above.

Example 3.41

More interestingly, there is a similar map $s : S^3 \to SO(3)$, defined by

$$
s(w, x, y, z) = \begin{bmatrix} w^2 + x^2 - y^2 - z^2 & 2(xy - wz) & 2(wy + xz) \\ 2(xy + wz) & w^2 - x^2 + y^2 - z^2 & 2(yz - wx) \\ 2(xz - wy) & 2(yz + wx) & w^2 - x^2 - y^2 + z^2 \end{bmatrix}.
$$

As before, this is the restriction of a function $\mathbf{R}^4 \to \mathbf{R}^9$ which is continuous, being a composite of additions and multiplications. Once you have verified that $s(w, x, y, z)$ is in $SO(3)$ whenever (w, x, y, z) is in S^3, Proposition 3.35 assures us that s is continuous.

We can express any point of S^3 as $(\cos\theta, x\sin\theta, y\sin\theta, z\sin\theta)$, where $(x, y, z) \in S^2$ (Exercise 3.10). Applying s to this gives a matrix which rotates points by 2θ clockwise about the line through the point (x, y, z). (Again, you should confirm for yourself that this geometric description matches the algebraic formula given above.) Since all elements of $SO(3)$ correspond to rotations, this shows that s is surjective.

The map s is not injective, but it nearly is: If $s(w, x, y, z) = s(w', x', y', z')$, then either $(w, x, y, z) = (w', x', y', z')$ or $(w, x, y, z) = -(w', x', y', z')$. We will see later how this can help us understand the topology of $SO(3)$ in terms of S^3.

3.4 Bases

Proving continuity directly can be quite awkward, which is why it is convenient to use shortcuts such as showing that the function is a composite of two continuous functions, or a restriction of a continuous function. An alternative approach is to reduce the number of open sets that we need to look at.

Recall that the definition of open subset of $\mathbf{R}^n$ says that around every point in the subset there is an open ball contained in the subset. So the subset is the union of all these open balls. In other words, every open subset is a union of open balls. We say that the open balls form a "basis"[2] for the topology on $\mathbf{R}^n$.

Definition: In a topological space T, a collection $\mathcal{B}$ of open subsets of T is said to form a **basis** for the topology on T if every open subset of T can be written as a union of sets in $\mathcal{B}$.

Example 3.42

The collection of all finite open intervals (a, b) (where $a, b \in \mathbf{R}$) forms a basis for the topology on $\mathbf{R}$, since every open set contains an open interval around each point in it and, hence, is the union of all these open intervals.

Example 3.43

If the set $\{1, 2, 3\}$ is given the discrete topology, so that every subset is open, then the sets $\{1\}$, $\{2\}$, $\{3\}$ form a basis for this topology, as every subset of $\{1, 2, 3\}$ can be written as a union of some of these basic open sets.

If we have a basis for the range space of a given function, then we need only check the preimages of those basic open sets in order to verify continuity of the function, thanks to the following result.

---

[2] Note that this use of the word "basis" has no connection with the usage of the word "basis" in linear algebra.

Proposition 3.44

If $f : S \to T$ is a function between two topological spaces S and T, and T has a basis $\mathcal{B}$, then f is continuous if $f^{-1}(B)$ is open for every set B in the basis $\mathcal{B}$.

Proof

For f to be continuous we need to show that $f^{-1}(Q)$ is open for any open set $Q \subset T$. Now, as $\mathcal{B}$ is a basis for the topology on T, we can write Q as a union of some sets in $\mathcal{B}$. Then the preimage $f^{-1}(Q)$ of Q is the union of the preimages of these basic sets, by Lemma 2.8. Since the preimage of every set in $\mathcal{B}$ is open, so $f^{-1}(Q)$ is a union of open sets, and hence open. Thus f is continuous. □

Example 3.45

Let $f : \mathbf{R} \to \mathbf{R}$ be the function $f(x) = 2x + 3$ of Example 2.10. A basis for the topology on $\mathbf{R}$ is given by the collection of all open intervals (a, b). Since f is an increasing function, $f(x) \in (a, b)$ if, and only if, $x \in (\frac{a-3}{2}, \frac{b-3}{2})$. So $f^{-1}(a, b) = (\frac{a-3}{2}, \frac{b-3}{2})$ which is open. Hence f is continuous.

Example 3.46

Let $f : \mathbf{R} \to \mathbf{R}$ be the function $f(x) = x^2$. Again we use the basis of open intervals (a, b); their preimages are as follows:

$$f^{-1}(a, b) = \begin{cases} (-\sqrt{b}, -\sqrt{a}) \cup (\sqrt{a}, \sqrt{b}) & \text{if} \quad 0 \le a < b, \\ (-\sqrt{b}, \sqrt{b}) & \text{if} \quad a < 0 < b, \\ \emptyset & \text{if} \quad b \le 0. \end{cases}$$

In each of these cases, the preimage is open, hence f is continuous.

Example 3.47

Let $e : \mathbf{R} \to S^1$ be the **exponential map** $e(x) = (\cos(2\pi x), \sin(2\pi x))$. This coils the real line anti-clockwise around the circle

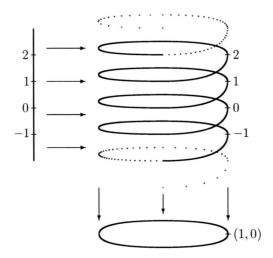

So, for example, $e(0) = (1,0)$, $e\left(\frac{1}{4}\right) = (0,1)$, $e\left(\frac{1}{2}\right) = (-1,0)$, $e\left(\frac{3}{4}\right) = (0,-1)$ and $e(1) = (1,0)$.

As S^1 is a subspace of $\mathbf{R}^2$, we can obtain a basis for S^1 from the basis for $\mathbf{R}^2$ of open balls. Every open set in S^1 is the intersection of S^1 with an open set in $\mathbf{R}^2$, and hence every open set in S^1 is a union of intersections of S^1 with open balls in $\mathbf{R}^2$.

Most of these intersections will either be empty or will contain all of S^1. The preimage of such intersections will then be either empty or the whole of $\mathbf{R}$. In either case, the preimage will be open.

All other intersections will have the form of open arcs, such as $S^1 \cap B_1(1,1)$ which is depicted in Example 3.19. The preimage of this arc is an infinite union

$$\cdots (-2, \frac{-7}{4}) \cup (-1, \frac{-3}{4}) \cup (0, \frac{1}{4}) \cup (1, \frac{5}{4}) \cup \cdots$$

of open intervals, hence it is an open subset of $\mathbf{R}$.

Similarly, the preimage of any open arc will be a union of open intervals, one for each integer, so all such preimages will be open. Thus the exponential map e is continuous.

EXERCISES

3.1. How many topologies can be put on: 1) a set that has 2 points? 2) a set that has 3 points? 3) a set that has 4 points?

3.2. Let $S = \{a, b, c, d\}$. Which of the following lists of "open" sets forms a topology on S (i.e., which lists satisfy the axioms T1–T4)?

- $\emptyset$, $\{a\}$, $\{a,b\}$, $\{a,b,c,d\}$
- $\emptyset$, $\{a\}$, $\{b\}$, $\{a,b,c,d\}$
- $\emptyset$, $\{a,c\}$, $\{a,b,c\}$, $\{a,c,d\}$, $\{a,b,c,d\}$
- $\{a\}$, $\{a,b\}$, $\{a,b,c\}$, $\{a,b,c,d\}$

3.3. For each topology on $\{a,b,c,d\}$ from Question 3.2, list the open sets in the subspace topology for the subset $\{a,b,c\}$.

3.4. Prove that a function $f : S \to T$ between two topological spaces is continuous if, and only if, $f^{-1}(C)$ is closed whenever $C \subset T$ is closed.

3.5. Verify that the open sets of $\mathbf{R}^2$ defined in Example 3.11 do form a topology on $\mathbf{R}^2$.

3.6. Let $m : \mathbf{R}^2 \to \mathbf{R}$ be the multiplication function $m(x,y) = xy$. Sketch the preimage of the open interval $(1,2)$ and show that this preimage is open.

3.7. Verify that the subspace topology on a subset S of a topological space T is, in fact, a topology.

3.8. Let $f : \mathbf{R} \to \mathbf{Z}$ be the "floor" function which rounds a real number x down to the nearest integer:

$$f(x) = n \text{ provided that } n \in \mathbf{Z} \text{ and } n \le x < n+1.$$

Determine whether or not f is continuous.

3.9. Let $\mathbf{R}^2 - \mathbf{R}$ be the subset of $\mathbf{R}^2$ consisting of all pairs (x,y) with $y \ne 0$. Define $d : \mathbf{R}^2 - \mathbf{R} \to \mathbf{R}$ to be the division function $d(x,y) = x/y$. Describe the preimage $d^{-1}(a,b)$ of an arbitrary open interval (a,b). Determine whether or not d is continuous.

3.10. Verify that every point $(w,x,y,z) \in \mathbf{R}^4$ satisfying $w^2+x^2+y^2+z^2 = 1$ can be expressed as $(\cos\theta, x'\sin\theta, y'\sin\theta, z'\sin\theta)$ where $0 \le \theta < 2\pi$ and $(x',y',z') \in S^2$.

3.11. Let T be a set and B a collection of subsets of T. Show that if every element of T belongs to at least one subset in B and B is closed under finite intersections, then the collection of all unions of sets in B forms a topology on T.

Interlude

We have now seen enough examples of topological spaces and continuous maps to give us a flavour of the basic landscape of topology. In the rest of the book we will study the features of this landscape. Our ultimate goal is always to use the topology of a space to get as much information as possible about continuous functions from, or to, that space.

Most of the tools that we use in studying topology can be grouped into two categories: Topological constructions and topological invariants.

A topological construction takes one or more topological spaces and builds a new space out of them. For example, in Chapter 5, we will see how to take two spaces S and T and form their "product" $S \times T$ which, if $S = T = \mathbf{R}$, gives $\mathbf{R} \times \mathbf{R} = \mathbf{R}^2$, the Euclidean plane of Example 3.11. If we are faced with a new and unfamiliar space that we wish to understand, then we can try to "deconstruct it" by giving a recipe for how to construct the new space from other spaces that we are more familiar with. For example, we may be able to express the new space as the product of two other spaces, in which case we can use our knowledge of these spaces to deduce information about the new space. Clearly, the more constructions that we know of, the more successful this deconstructionist approach will be. So, in Chapter 5, we will meet three methods of construction, as well as considering how to tell if the new space is actually identical, topologically speaking, with a space we are familiar with. As we will see, familiar spaces can take on many different forms, so we need a way of seeing through such disguises.

A topological invariant is any characteristic of the space which it shares with all topologically-identical spaces. The simplest examples of these are "topological properties", some of which we will meet in Chapter 4. For example, we will learn about the property of "connectivity" which is such that a connected space cannot map surjectively onto a disconnected space in a continuous way.

Thus, as soon as we can show some spaces to be connected and others to be disconnected, we will obtain several theorems about continuous functions. For example, using this approach we will prove that every continuous integer-valued function on the real line is constant.

In fact, the remainder of the book is concerned with these two types of tools: Chapters 6–10 with more subtle topological invariants, and Chapter 11 with more topological constructions.

4
Topological Properties

4.1 Connectivity

A typical example of the type of statement about continuous maps that topologists try to prove is the following.

Theorem 4.1

There is no continuous surjective map $\mathbf{R} \to S^0$.

Proof

Suppose that $f : \mathbf{R} \to S^0$ is such a map. Since S^0 is discrete, the subsets $\{-1\}$ and $\{+1\}$ are open and, if f is continuous, then the subsets $U = f^{-1}\{-1\}$ and $V = f^{-1}\{+1\}$ of $\mathbf{R}$ are open. Since $\{-1\}$ and $\{+1\}$ constitute, between them, the whole of S^0, so $U \cup V$ must be the whole of $\mathbf{R}$. As f is surjective, both U and V contain at least one point each; they are non-empty. And, because $\{-1\}$ and $\{+1\}$ are disjoint, so $U \cap V = \emptyset$. We will see that there can be no such open subsets of $\mathbf{R}$.

To do this, let x be any point in U and let y be any point in V. By swapping U and V if necessary, we can assume that $x < y$. Thus we have an interval $[x, y]$ with one end point in U and one in V. The midpoint $\frac{x+y}{2}$ is either in U or in V. If it is in V, then the interval $[x, \frac{x+y}{2}]$ has one end point in U and one end point in V. On the other hand, if $\frac{x+y}{2}$ is in U, then the interval $[\frac{x+y}{2}, y]$

M.D. Crossley, *Essential Topology*, Springer Undergraduate
Mathematics Series, DOI 10.1007/978-1-84628-194-5_4,
© Springer-Verlag London Limited 2010

has one end point in U and one end point in V. Either way, we can produce a closed interval of length $\frac{y-x}{2}$, with one end point from each of the sets U and V. Similarly, we can cut this interval in half and one of the two halves will be an interval of length $\frac{y-x}{4}$ with one end point from each set. Carrying on, we can produce intervals of decreasing length with one end in either set. The intersection of all of these intervals will be a single point z of $\mathbf{R}$. If this point lies in U, then there will be an open interval $(z - \delta_z, z + \delta_z)$ contained in U for some $\delta_z > 0$. Then any interval of length $< \delta_z$ containing z will be contained in $(z - \delta_z, z + \delta_z)$ and hence in U. In particular, any interval of length $\frac{y-x}{2^n}$ that contains z will be contained in U if n is large enough (n must be greater than $\log_2(\frac{y-x}{\delta_z})$). But we know there is such an interval with one end point in U and one in V. One of these end points then lies both in U and V which cannot happen since $U \cap V = \emptyset$. Hence z cannot lie in U. Exactly the same argument can be applied if z lies in V, so we have a contradiction.

The only answer must be that the points x and y cannot have existed, i.e., either U is empty or V is empty.   □

This proof relies on the fact that two open subsets of $\mathbf{R}$ cannot be disjoint, cover the whole of $\mathbf{R}$ and both be non-empty. A space, like $\mathbf{R}$, with this property is said to be 'connected':

Definition: We say that a topological space T is **disconnected** if it is possible to find two open subsets U, V of T such that:

– U and V have no intersection, i.e., $U \cap V = \emptyset$

– U and V cover T, i.e., $U \cup V = T$

– Neither U nor V are empty, i.e., $U \neq \emptyset$ and $V \neq \emptyset$.

If there are no such subsets within T, then T is **connected**.

Example 4.2

The space $\mathbf{R}$ is connected. This is what the second part of the proof of Theorem 4.1 shows.

The first part of the proof of Theorem 4.1 then tells us the following.

Lemma 4.3

If T is connected, then there is no continuous surjection $T \to S^0$.

Example 4.4

The open interval $(0, 1)$ is connected. This can be proved in exactly the same way as for $\mathbf{R}$. Hence there are no continuous surjective maps $(0, 1) \to S^0$.

Example 4.5

Similarly, a closed interval such as $[0, 1]$ is connected, as can be seen using the same proof again.

Combining this with Lemma 4.3 and a little trick yields the following fixed-point theorem.

Theorem 4.6 (Fixed-Point Theorem for $[0, 1]$)

If $f : [0, 1] \to [0, 1]$ is a continuous map, then f has a fixed point, i.e., there is some $x \in [0, 1]$ such that $f(x) = x$.

Proof

Suppose that $f : [0, 1] \to [0, 1]$ is continuous but has no fixed points. We can define a map g by

$$g(x) = \frac{x - f(x)}{|x - f(x)|}$$

for $x \in [0, 1]$. Since $f(x) \neq x$, g is a composite of additions and divisions by non-zero numbers, so it is continuous. The value of $g(x)$ is either $+1$ or -1, so we can think of g as a map $[0, 1] \to S^0$ which, by Proposition 3.35, is continuous. Now $f(0) \in [0, 1]$ and $f(0) \neq 0$, so $f(0) > 0$, i.e., $g(0) = -1$. Similarly, $f(1) < 1$ so $g(1) = +1$. Thus g gives a surjective continuous map $[0, 1] \to S^0$. By Example 4.5 and Lemma 4.3, no such map can exist. $\qquad\square$

Another consequence of Lemma 4.3 is the following proof of the intermediate value theorem:

Theorem 4.7 (Intermediate Value Theorem)

If $f : [a, b] \to \mathbf{R}$ is a continuous function and $f(a) < 0$ and $f(b) > 0$, then there is at least one point $x \in [a, b]$ such that $f(x) = 0$.

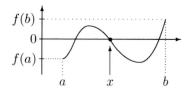

Proof

Suppose that there is no such x, i.e., $f(x) \neq 0$ for all $x \in [a,b]$. Then we can define a new map $\bar{f} : [a,b] \rightarrow S^0$ by $\bar{f}(x) = f(x)/|f(x)|$. This is continuous as f is, and surjective since $f(a) < 0$ so $\bar{f}(a) = -1$, and $f(b) > 0$ so $\bar{f}(b) = +1$.

The interval $[a,b]$ is connected (the argument of Example 4.5 can easily be adapted to show this) so, by Lemma 4.3, there can be no such continuous map. Hence our assumption that $f(x) \neq 0$ for all $x \in [a,b]$ must have been incorrect, i.e., there must be some x for which $f(x) = 0$. $\qquad\square$

Example 4.8

The proof that $\mathbf{R}$ is connected can also be used to show that $\mathbf{R}^n$ is connected.

Example 4.9

With a little more care, the argument can also be used to show that $\mathbf{R}^n - \{0\}$ is connected if $n > 1$. As usual, we start by supposing that $\mathbf{R}^n - \{0\} = U \cup V$, where U and V are non-empty and disjoint open sets. Then we take a point $x \in U$ and $y \in V$. If the straight line from x to y is contained in $\mathbf{R}^n - \{0\}$ (i.e., this line does not pass through 0), then we can proceed as usual.

If, however, the line from x to y passes through 0, then we need to replace one of the end points. To do this, take any point in $\mathbf{R}^n$ which is not on the straight line through x and y. This point will either lie in U or in V. If it is in U, then we replace x by this new point and, if it is in V, then we replace y by it. Either way we end up with two points, one in U and one in V, such that the straight line between the two points does not pass through 0, so is contained in $\mathbf{R}^n - \{0\}$. Then the rest of the proof goes through unchanged.

If $n = 1$, then we cannot use this argument and, in fact, $\mathbf{R} - \{0\}$ is disconnected, as we can define $U = (0, \infty)$ and $V = (-\infty, 0)$.

Lemma 4.3 also has a converse.

Lemma 4.10

If T is disconnected, then there is a continuous surjection $T \to S^0$.

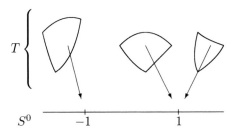

Proof

If T is disconnected, then there are non-empty open subsets $U, V \subset T$ with $U \cap V = \emptyset$, $U \cup V = T$. We define $f : T \to S^0$ as follows. As U and V are disjoint, we can define $f(x) = 1$ for $x \in U$ and $f(x) = -1$ for $x \in V$. Since $U \cup V = T$, this defines f completely, and f is clearly surjective. The preimage of $\{1\} \subset S^0$ is U and the preimage of $\{-1\} \subset S^0$ is V, so both of these are open, hence f is continuous.   □

This implies the following.

Proposition 4.11

If S is a connected space and T is a disconnected space, then there can be no surjective continuous map $S \to T$.

Proof

If T is disconnected, then there is a continuous surjection $T \to S^0$. If there is also a continuous surjection $S \to T$, then we can combine these to get a continuous surjection $S \to S^0$ which cannot happen if S is connected.   □

Example 4.12

There is no continuous surjection $\mathbf{R} \to \mathbf{R} - \{0\}$ since $\mathbf{R}$ is connected and $\mathbf{R} - \{0\}$ is not.

Proposition 4.11 can also be used both to prove that spaces are connected and to prove that they are not. For example:

Example 4.13

The circle S^1 is connected, because $\mathbf{R}$ is connected and the map $e : \mathbf{R} \to S^1$ of Example 3.47 is surjective and continuous.

Example 4.14

The n-sphere S^n is connected if $n > 0$, because we can define a continuous surjection $\mathbf{R}^{n+1} - \{0\} \to S^n$ by $x \mapsto x/|x|$, depicted below for $n = 1$.

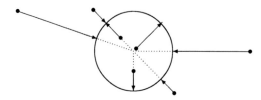

Example 4.15

The projective space $\mathbf{RP}^n$ is connected if $n > 0$, because there is a continuous surjection $\mathbf{R}^{n+1} - \{0\} \to \mathbf{RP}^n$ which takes a point in $\mathbf{R}^{n+1} - \{0\}$ to the line through the origin containing this point.

Example 4.16

The space $\mathrm{GL}(3, \mathbf{R})$ is disconnected, because $\mathbf{R} - \{0\}$ is disconnected and there is a continuous surjection $\mathrm{GL}(3, \mathbf{R}) \to \mathbf{R} - \{0\}$ defined by

$$M \mapsto \det(M).$$

(This is surjective, because if $\lambda \in \mathbf{R} - \{0\}$, then the matrix

$$\begin{bmatrix} \lambda & 0 & 0 \\ 0 & 1 & 0 \\ 0 & 0 & 1 \end{bmatrix}$$

is invertible and has determinant λ.)

Similarly, the space $\mathrm{O}(3)$ is disconnected, because det restricts to a surjection $\mathrm{O}(3) \to S^0$. On the other hand, $\mathrm{SO}(3)$ is connected:

Example 4.17

The space SO(3) is connected because in Example 3.41 we met a continuous surjection $S^3 \to$ SO(3) which, because S^3 is connected, can only exist if SO(3) is connected.

We can also develop the ideas of Theorem 4.1 in the following way.

Lemma 4.18

If S is a connected space and T is a discrete space, then any continuous map $f : S \to T$ is constant.

Proof

Let u be a point in the image of f, say $u = f(x)$ for some $x \in S$. The set $\{u\}$ is open as T is discrete. And $T - \{u\}$, the set of all points in T other than u, will also be open. Hence $f^{-1}(\{u\})$ and $f^{-1}(T - \{u\})$ will both be open. Moreover, they will be disjoint, since u and $T - \{u\}$ are disjoint. And they will cover S as $\{u\} \cup (T - \{u\}) = T$. Since S is connected, either $f^{-1}(\{u\})$ or $f^{-1}(T - \{u\})$ will be empty. The preimage $f^{-1}(\{u\})$ cannot be empty as it contains x. Hence it must be the case that $f^{-1}(T - \{u\})$ is empty, and $f^{-1}\{u\} = S$. This says that the image of f is just $\{u\}$, i.e., f is constant. □

Since $\mathbf{R}$ is connected and $\mathbf{Z}$ is discrete, this implies the following familiar result:

Corollary 4.19

Every continuous map from $\mathbf{R}$ to $\mathbf{Z}$ is constant.

4.2 Compactness

Another typical example of a topological theorem is that there is no continuous surjection $[0, 1] \to \mathbf{R}$. In fact we can say more: Every continuous function $[0, 1] \to \mathbf{R}$ is "bounded":

Proposition 4.20

If $f : [0, 1] \to \mathbf{R}$ is a continuous function, then f is **bounded**, i.e., there are numbers j, k such that Im $f \subset (j, k)$. In other words, $j < f(x) < k$ for all $x \in [0, 1]$.

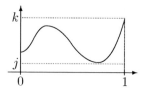

Partial Proof

To prove this, suppose that $f : [0, 1] \to \mathbf{R}$ is a continuous function. Consider the overlapping open intervals $\dots (-2, 0), (-1, 1), (0, 2), (1, 3), \dots$ in $\mathbf{R}$. These form an **open cover** of $\mathbf{R}$, meaning that every point in $\mathbf{R}$ is contained in at least one of these open intervals.

Since f is continuous, the preimage of each interval $(i, i + 2)$ is an open subset of $[0, 1]$. Every point in $[0, 1]$ must get mapped by f into one of these intervals $(i, i + 2)$ and so must belong to one of the preimages $f^{-1}(i, i + 2)$. Hence these preimages $f^{-1}(i, i + 2)$ (taken over all integers i) form an open cover of $[0, 1]$.

If f is bounded, then its image is contained in a finite union of these intervals: If $j < f(x) < k$ for all x, then

$$\text{Im } f \subset (j, j + 2) \cup (j + 1, j + 3) \cup \cdots \cup (k - 2, k).$$

Putting this another way, the preimages

$$f^{-1}(j, j + 2), f^{-1}(j + 1, j + 3), \dots, f^{-1}(k - 2, k)$$

must cover $[0, 1]$. So, if f is bounded, then, of our original open cover consisting of all the preimages $f^{-1}(i, i + 2)$, we can discard all but finitely many and still have a cover of $[0, 1]$.

The converse is also true. If there is a finite number of preimages $f^{-1}(i_1, i_1 + 2), \dots, f^{-1}(i_n, i_n + 2)$ which cover $[0, 1]$, then the image of f is contained in the union

$$(i_1, i_1 + 2) \cup (i_2, i_2 + 2) \cup \cdots \cup (i_n, i_n + 2).$$

Hence Im $f \subset (j, k)$, where $j = \min(i_1, \dots, i_n)$, $k = \max(i_1 + 2, \dots, i_n + 2)$.

So if we can prove that any open cover of $[0, 1]$, such as $\{f^{-1}(i, i+2), \ i \in \mathbf{Z}\}$ can be "refined" to a finite open cover, then we will be able to deduce that any continuous function $f : [0, 1] \to \mathbf{R}$ is bounded.

To complete this proof, then, we need to know that given a certain covering of $[0,1]$ by an infinite number of open sets, it is possible to throw away all but a finite number of these open sets and still cover $[0,1]$. In other words, we need to know that the infinite open covering of $[0,1]$ has a "finite refinement".

The open covering of $[0,1]$ that occurred in the proof was formed by preimages $f^{-1}(i, i+2)$, where i is any integer. These preimages change if we change f, and they can vary greatly. So we need to know that *every* open cover of $[0,1]$ has a finite refinement. This property is called "compactness".

Definition: An **open cover** of a topological space T is a collection of open subsets of T such that every point in T lies in at least one of these open subsets.

A topological space T is said to be **compact** if every open covering of T admits a finite refinement. In other words, given any infinite collection of open sets, which covers T, it is possible to throw most of these sets away, keeping only a finite number of them, and still have an open covering of T.

To complete the proof, then, we need to show that the interval $[0,1]$ is compact.

Proposition 4.21

The closed interval $[0,1]$ is compact.

Proof

To prove that $[0,1]$ is compact, suppose that we have an open cover of $[0,1]$. We will show that this open cover has a finite refinement by contradiction, so let us assume that there is no finite refinement of this cover. Now consider the intervals $[0,1/2]$ and $[1/2,1]$. Intersecting the original open cover with each of these intervals gives an open cover for each interval. If they both admit a finite refinement, then we can combine these to get a finite refinement of the original open cover. We have assumed that this is not the case, so one of the intervals must not have a finite refinement. Let I_1 be this interval.

Now we divide I_1 in two closed intervals: if $I_1 = [0,1/2]$, then we divide it up into $[0,1/4]$ and $[1/4,1/2]$, and if $I_1 = [1/2,1]$, then we use the intervals $[1/2,3/4]$ and $[3/4,1]$. The same argument shows that one of these two halves must not have a finite refinement; let I_2 be that half. We then divide I_2 in half, and so on. Carrying on in this way, we obtain a nested series of closed intervals:

$$[0,1] \supset I_1 \supset I_2 \supset \cdots$$

of decreasing length: I_n has length $1/2^n$, and none of these intervals admits a finite refinement of the open cover. The intersection of all of these intervals will be a single point $c \in [0, 1]$. This must be contained in one of the sets in the open cover. Since this set is open, it must also contain some breathing space $(c - \delta_c, c + \delta_c) \cap [0, 1]$, where $\delta_c > 0$. But this will contain any interval around c in $[0, 1]$ if the length of the interval is less than δ_c. In particular, $(c - \delta_c, c + \delta_c)$ will contain I_n if $n > \log_2(1/\delta_c)$. So, for n large enough, the interval I_n will be contained in $(c - \delta_c, c + \delta_c)$ which, we noted, was contained in one of the open sets in the cover. Hence I_n has a finite refinement of the original cover, in fact consisting of just one set. But the intervals I_n were chosen so that none of them admits a finite refinement. Thus we have a contradiction, showing that our assumption, that the interval $[0, 1]$ has no finite refinement, must have been wrong. In other words, there *is* such a finite refinement. Thus $[0, 1]$ is compact. □

So $[0, 1]$ is compact, and the proof of Proposition 4.20 is complete. Now, notice that the proof of Proposition 4.20 didn't use any information about $[0, 1]$ *apart* from the fact that it was compact. So the same proof can be used for the following.

Proposition 4.22

If T is a compact topological space and $f : T \to \mathbf{R}$ is a continuous function, then f is bounded.

In other words, all spaces which are compact have this property about continuous functions to $\mathbf{R}$. If we can show that any given space is compact, we will be able to deduce this property about real-valued functions on that space.

Lemma 4.23

The circle S^1 is compact.

Proof

We will prove this by relating S^1 to the interval $[0, 1]$ which we know is compact. Let $e : [0, 1] \to S^1$ be the map $e(t) = (\cos(2\pi t), \sin(2\pi t))$, the restriction of the exponential map of Example 3.47.

Now, let $\mathcal{U}$ be any open cover of S^1, so $\mathcal{U}$ is a collection of open subsets of S^1. For each subset $Q \in \mathcal{U}$, we have an open subset $f^{-1}(Q)$ of $[0, 1]$, as f is continuous. The collection of all of these open subsets $f^{-1}(Q)$ for $Q \in \mathcal{U}$ is an open cover of $[0, 1]$, and so has a finite refinement. Let $\mathcal{V}$ be a finite collection of open subsets Q taken from $\mathcal{U}$ such that $\cup_{Q \in \mathcal{V}} f^{-1}(Q)$ covers $[0, 1]$. Then $\mathcal{V}$ is a finite refinement of $\mathcal{U}$ for the circle S^1. For if $x \in S^1$, then x is in the image of f, as f is surjective. Hence there is some $y \in [0, 1]$ such that $f(y) = x$. As $[0, 1]$ is covered by $\cup_{Q \in \mathcal{V}} f^{-1}(Q)$, there must be some $Q \in \mathcal{V}$ such that $y \in f^{-1}(Q)$. If $y \in f^{-1}(Q)$, then $f(y) \in Q$, so $x \in Q$. Hence, for each point x in S^1, there is some subset in $\mathcal{V}$ which contains x. Thus $\mathcal{V}$ is a finite refinement of $\mathcal{U}$, and S^1 is compact. □

Having proved that, we get the following corollary for nothing.

Corollary 4.24

Any continuous map from S^1 to $\mathbf{R}$ is bounded.

Of course, many spaces are not compact.

Example 4.25

The space $\mathbf{R}$ cannot be compact because not all continuous functions from $\mathbf{R}$ to $\mathbf{R}$ are bounded (for example, the identity function $f(x) = x$ is not bounded on $\mathbf{R}$).

One open cover of $\mathbf{R}$ that does not have a finite refinement is the following. Let $U_n = (-n, n)$ for each positive integer n. Every real number belongs to one of these open intervals, so they form an open cover of $\mathbf{R}$. However, if we take a finite number of these intervals, say $U_{i_1}, U_{i_2}, \ldots, U_{i_k}$, then these will not cover $\mathbf{R}$. For their union will simply equal $(-i, i) = U_i$, where $i = \max(i_1, \ldots, i_k)$. There are many real numbers which are not contained in U_i, so any such finite refinement will fail to be an open cover of $\mathbf{R}$. Hence the original, infinite open cover cannot be refined finitely to give an open cover.

Example 4.26

One can show that the open interval $(0, 1)$ is not compact in a similar way, defining open sets U_n for all integers $n > 1$ by

$$U_2 = (\frac{1}{2}, 1), U_3 = (\frac{1}{3}, 1), \ldots, U_n = (\frac{1}{n}, 1), \ldots.$$

These cover $(0, 1)$, because if any real number x is between 0 and 1 then it is also between $1/n$ and 1 for some sufficiently large integer n. But if we take any finite refinement, say $U_{i_1}, \ldots, U_{i_k}$, then this will not cover $(0, 1)$. For, the union of such a finite collection will again be U_i, where i is the maximum of $i_1, \ldots, i_k$. And U_i omits some points from $(0,1)$, such as $1/i$. Hence no finite refinement of this cover is itself a cover of $(0,1)$ and so $(0,1)$ is not compact.

The proof of Lemma 4.23 can be very easily generalized to prove the following result.

Proposition 4.27

If $f : S \to T$ is a continuous map and S is compact, then the image of f is compact.

That is to say, the image of a compact space under a continuous map is compact. Combining this with Example 4.26 gives a slightly surprising result.

Corollary 4.28

There is no continuous surjective map $[0,1] \to (0,1)$.

This demonstrates the power of knowing whether a space is compact or not. Fortunately, for subspaces of $\mathbf{R}^n$, we can say precisely which are compact and which are not. This is expressed in the following theorem, which relates compactness to the familiar property of being closed (in the topology on $\mathbf{R}^n$), and the simple property of being "bounded". We say that a subset S of $\mathbf{R}^n$ is **bounded** if there is some $x \in \mathbf{R}^n$ and finite radius $\delta > 0$ such that S is contained in the open ball $B_\delta(x)$.

Theorem 4.29 (Heine–Borel Theorem)

A subspace T of $\mathbf{R}^n$ is compact if, and only if, T is closed (as a subset of $\mathbf{R}^n$) and bounded.

Proof

For clarity, we will only prove this in the case $n = 1$, the general case being very similar. Let T be a closed, bounded subspace of $\mathbf{R}$. The fact that T is bounded means that there are two numbers $a, b \in \mathbf{R}$ such that $a < x < b$ whenever $x \in T$. If T is closed, then its complement $\mathbf{R} - T$ is open. If we have an open cover of T, then, by definition of the subspace topology, each open set is the intersection of T with an open subset of $\mathbf{R}$. Taking all these open subsets of $\mathbf{R}$ together with the open set $\mathbf{R} - T$ must give an open cover of $\mathbf{R}$. Intersecting all these open subsets of $\mathbf{R}$ (including $\mathbf{R} - T$) with $[a, b]$ will then give an open cover of $[a, b]$.

Proposition 4.21 showed that $[0, 1]$ is compact, and the argument can be easily modified to show that any closed interval $[a, b] \subset \mathbf{R}$ is compact. Hence the open cover of $[a, b]$ that we have just constructed must admit a finite refinement. So we can discard all but finitely many of the open subsets of $\mathbf{R}$ and still have a cover of $[a, b]$ when we intersect with this interval. If this finite list of open subsets covers $[a, b]$, then it will certainly cover T. By intersecting these open subsets of $\mathbf{R}$ with T we will have a finite refinement of the original cover of T, since the only extra set we added was $\mathbf{R} - T$, and this will vanish when we intersect with T, since $\mathbf{R} - T \cap T = \emptyset$. Hence T is compact.

On the other hand, suppose that T is not a closed set of $\mathbf{R}$. So the complement of T is not open, and therefore has some point $x \notin T$ such that no neighbourhood of x is contained in the complement of T. In other words, every interval $(x - \delta, x + \delta)$, with $\delta > 0$, contains an element of T.

Let I_n be the intersection of T with the complement of the interval $[x - \frac{1}{n}, x + \frac{1}{n}]$. So $y \in I_n$ if $y \in T$ and either $y < x - \frac{1}{n}$ or $y > x + \frac{1}{n}$.

If now y is any element of T, then $y \neq x$ so $|y - x| > 0$. In particular, we can find an integer n such that $|y - x| > 1/n$. Thus $y \notin [x - \frac{1}{n}, x + \frac{1}{n}]$. Since $y \in T$, we see that $y \in I_n$. Hence the union of the I_n's contains every element of T. So this is an open cover of T.

Now $I_1 \subset I_2 \subset I_3 \subset \cdots$ and so any finite union of I_n's will be equal to I_m, where m is the largest index involved. And I_m does not contain $[x - \frac{1}{m}, x + \frac{1}{m}]$. But every interval $(x - \frac{1}{m}, x + \frac{1}{m})$ contains a point of T. Hence I_m cannot cover T, and so there can be no finite refinement of the I_n's which covers T. Thus T is not compact.

Finally, suppose that T is not bounded. Then let $I_n = (-n, n)$ for each integer $n > 0$. Every real number x is contained in I_n for some n. So the I_n's cover $\mathbf{R}$ and, consequently, T.

If we take a finite number of these I_n's, then their union will be equal to I_m for some m. If T is not bounded, then for each real number k, there is some $x \in T$ such that either $x > k$ or $x < -k$. In particular, for every integer $n > 0$, there is some $x \in T$ such that $x \notin I_n$. Hence I_m cannot cover T, so T is not compact. $\qquad\square$

Corollary 4.30

The sphere S^n is compact.

4.3 The Hausdorff Property

The last property that we will meet for now is the "Hausdorff" property.

> Definition: We say that a topological space T is **Hausdorff** if, for any two distinct points x, y in T, there are open subsets U, V of T such that $x \in U$ and $y \in V$ and $U \cap V = \emptyset$.

In other words, around any two distinct points, we can find two non-overlapping open sets, as depicted below.

The traditional way of saying this is that any two distinct points can be "housed off" from each other.

Example 4.31

The interval $[0, 1]$ is Hausdorff. To prove this, let $x, y \in [0, 1]$ be two distinct points. Then $|y-x| > 0$, and we can set $\delta = |y-x|/2$. Let $U = (x-\delta, x+\delta) \cap [0, 1]$ and $V = (y-\delta, y+\delta) \cap [0, 1]$. As these are intersections of $[0, 1]$ with open sets, they are both open. And $x \in U$, $y \in V$ and $U \cap V = \emptyset$, by the way we chose δ.

Example 4.32

Similarly, $\mathbf{R}^n$ is Hausdorff, for any n.

If a space is Hausdorff, then we have the following information about self-maps of the space:

Proposition 4.33

If T is a Hausdorff space and $f : T \to T$ is a continuous map, then the **fixed-point set**

$$\text{Fix}(f) = \{x \in T : f(x) = x\}$$

is a closed subset of T.

Proof

To prove that a subset is closed we must prove that its complement is open.

So let y be a point in the complement of Fix(f). Then as y is not a fixed point of f, we see that $f(y) \neq y$. Thus we have two distinct points $y, f(y)$ of T, so we can find open sets U, V, with $y \in U$, $f(y) \in V$ and $U \cap V = \emptyset$. Since f is continuous, $f^{-1}(V)$ is open, and so $U \cap f^{-1}(V)$ will be an open set containing y. Moreover, $U \cap f^{-1}(V)$ is disjoint from Fix(f). For suppose that $x \in U \cap f^{-1}(V)$ and $f(x) = x$. If $x \in U \cap f^{-1}(V)$, then $x \in f^{-1}(V)$ and so $f(x) \in V$. But also $x \in U$ and if $f(x) = x$, then $f(x) \in U$. So we have $f(x) \in U \cap V$, whereas $U \cap V$ is empty. So $U \cap f^{-1}(V)$ is disjoint from Fix(f) as claimed.

Hence around every point y in the complement of Fix(f), we can find an open set containing y and contained in the complement of Fix(f). The union of all these open sets (one for each point in the complement) will still be open, will be contained in the complement of Fix(f) and will also cover that complement. Hence the complement is open and Fix(f) is closed. □

Corollary 4.34

If $f : \mathbf{R} \to \mathbf{R}$ is a continuous map and $f(x) \neq x$, then there is an open interval $(x - \delta, x + \delta)$ of some positive radius δ containing no fixed points. In other words, if $y \in (x - \delta, x + \delta)$, then $f(y) \neq y$.

Proof

The set of points $x \in \mathbf{R}$ for which $f(x) \neq x$ is the complement of the set Fix(f). Since $\mathbf{R}$ is Hausdorff, this complement is closed, i.e., the set of non-fixed points is open. □

The following result tells us that most spaces that we meet will be Hausdorff:

Proposition 4.35

If $f : S \to T$ is continuous and injective and T is Hausdorff, then S is Hausdorff.

Proof

Let $x, y \in S$ be two distinct points. As f is injective, $f(x)$ and $f(y)$ in T will be distinct. Therefore, there are open subsets $U, V \subset T$ such that $f(x) \in U$, $f(y) \in V$ and $U \cap V = \emptyset$. Since f is continuous, the preimages $f^{-1}(U)$ and $f^{-1}(V)$ will be open subsets of S. Moreover, $x \in f^{-1}(U)$ and $y \in f^{-1}(V)$. Finally, $f^{-1}(U) \cap f^{-1}(V) = f^{-1}(U \cap V) = f^{-1}(\emptyset) = \emptyset$. □

So any subspace of a Hausdorff space is automatically Hausdorff. In particular, all subspaces of $\mathbf{R}^n$ are Hausdorff, so almost every space that we have met so far is Hausdorff.

Nevertheless, there are spaces which are not Hausdorff. A simple example can be obtained using the indiscrete topology.

Example 4.36

The set $\{1, 2\}$, with the indiscrete topology, is not Hausdorff. For if we take $x = 1$ and $y = 2$, then the only open set containing x is the whole set, which also contains y. Thus it is not possible to find two non-overlapping open sets each containing only one of the points.

A subtler example is the following.

Example 4.37

Let L be the real line together with an extra point which we'll call $0'$, and think of as an extra 0. We make this into a topological space in the following way. For every subset of $\mathbf{R}$ that does not contain 0, there is a corresponding subset of L, and we define this subset of L to be open if the corresponding subset of $\mathbf{R}$ is open. For each subset of $\mathbf{R}$ that contains 0, there are three corresponding subsets of L: One which contains 0, one which contains $0'$ and one which contains both. We define all three to be open if the original subset of $\mathbf{R}$ is open. The open sets on L that this gives do actually form a topology, i.e., they satisfy the axioms T1–T4, as you can check.

This construction gives a curious space called the **real line with a double point** at 0:

Most pairs of points in L can be "housed-off" from each other. However, if we take the points 0 and $0'$, then we cannot do this. For any open subset of L containing 0 or $0'$ corresponds to an open subset of $\mathbf{R}$ containing 0. Any two such open subsets of $\mathbf{R}$ will overlap, and their intersection will contain an interval $(-\delta, \delta)$. Hence, if we take an open subset of L containing 0 and one containing $0'$, then they will overlap. So L is not Hausdorff.

EXERCISES

4.1. Let S be a disconnected space, say $S = U \cup V$, where $U \cap V = \emptyset$ and U, V are both open and non-empty. If $x \in U$ and $y \in V$, prove that there can be no continuous map $f : [0,1] \to S$ such that $f(0) = x$ and $f(1) = y$.

4.2. Let T be the set $\{a, b, c\}$. For each of the following topologies on T, determine whether or not T is connected and whether or not T is Hausdorff:

- $\emptyset$, $\{a\}$, $\{a, b\}$, $\{a, b, c\}$.

- $\emptyset$, $\{a\}$, $\{b, c\}$, $\{a, b, c\}$.

- $\emptyset$, $\{a\}$, $\{a, b\}$, $\{a, c\}$, $\{c\}$, $\{a, b, c\}$.

4.3. Prove that every continuous map $\mathbf{R} \to \mathbf{Q}$ is constant. (Note that $\mathbf{Q}$ is not discrete, so Lemma 4.18 is not directly applicable.)

4.4. Prove that $\mathbf{R}^n$ is connected, as asserted by Example 4.8.

4.5. Determine whether or not the half-open interval $[0, 1/2)$ is compact. If it is, give a proof. If not, give an example of an unbounded function $[0, 1/2) \to \mathbf{R}$.

4.6. Using Proposition 4.27 and a proof by contradiction, show that $\mathrm{GL}(3, \mathbf{R})$ is not compact. What about $\mathrm{O}(3)$ and $\mathrm{SO}(3)$?

4.7. Prove that the set $\mathbf{Z}$ is Hausdorff without using Proposition 4.35.

4.8. Prove that if T is a Hausdorff space and $x_1, \ldots, x_n$ is a finite list of distinct points in T, then there are open sets $U_1, \ldots, U_n$ each containing one, and only one, of the points $x_1, \ldots, x_n$.

4.9. If the set of integers were given the indiscrete topology, would it be connected? Compact? Hausdorff?

4.10. Verify that the open sets on the real line with a double point, given in Example 4.37, do actually form a topology.

5
Deconstructionist Topology

In this chapter we consider ways of relating a new space to other spaces which are more familiar. Sometimes it will be the case that the new space is actually identical, topologically, with a space that we are more familiar with, and we will study this notion in Section 5.1. When this is not the case, it is often possible to express the new space in terms of other, perhaps more familiar, spaces. In the last three sections of this chapter, we will look at some different topological constructions which will enable us to deconstruct some of the more exotic spaces that we have met.

5.1 Homeomorphisms

For two spaces S and T to be identical topologically, they should be interchangeable in the sense that any continuous map with domain S has a corresponding continuous map with domain T, and vice versa. And any continuous map with range S should have a corresponding continuous map with range T, and vice versa. Since continuity is defined in terms of the topology, this is equivalent to requiring that there be a one-to-one correspondence between S and T which gives a one-to-one correspondence of their open sets. This can be phrased as follows.

M.D. Crossley, *Essential Topology*, Springer Undergraduate
Mathematics Series, DOI 10.1007/978-1-84628-194-5_5,

Definition: Two topological spaces S and T are said to be **homeomorphic** if there are continuous maps $f : S \to T$ and $g : T \to S$ such that

$$(f \circ g) = id_T \qquad \text{and} \qquad (g \circ f) = id_S.$$

The maps f and g are then **homeomorphisms**. These maps are inverse to each other, so we may write f^{-1} in place of g and g^{-1} in place of f. If S and T are homeomorphic, then we write $S \cong T$.

If we say that $f : S \to T$ is a homeomorphism, then we mean that f is continuous and there is a continuous map $g : T \to S$ inverse to f.

Example 5.1

Any two open intervals of the real line are homeomorphic. For example, if $S = (-1, 1)$ and $T = (0, 5)$, then define $f : S \to T$ and $g : T \to S$ by

$$f(x) = \frac{5}{2}(x + 1), \qquad g(x) = \frac{2}{5}x - 1.$$

These maps are continuous, being composites of addition and multiplication, and it is easy to verify that they are inverse to each other. So f and g are homeomorphisms, and $(-1, 1)$ and $(0, 5)$ are homeomorphic.

Example 5.2

If $S = \mathbf{R}$ and T is the open interval $(-1, 1)$, then S and T are homeomorphic. For we can define a continuous map $f : (-1, 1) \to \mathbf{R}$ by

$$f(x) = \tan(\frac{\pi x}{2}).$$

This is a bijection and has a continuous inverse $g : \mathbf{R} \to (-1, 1)$ given by

$$g(x) = \frac{2}{\pi} \arctan(x).$$

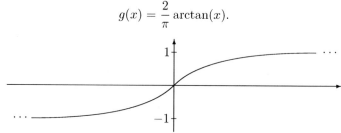

Graph of the function $g : \mathbf{R} \to (-1, 1)$ given by $g(x) = \frac{2}{\pi} \arctan(x)$.

As commented above, homeomorphic spaces are topologically identical. For example,

Proposition 5.3

If S is any topological space, then there is a one-to-one correspondence between continuous maps $(-1,1) \to S$ and continuous maps $\mathbf{R} \to S$. And there is a one-to-one correspondence between continuous maps $S \to (-1,1)$ and continuous maps $S \to \mathbf{R}$.

Proof

If $h : (-1,1) \to S$ is continuous, then $h \circ g : \mathbf{R} \to S$ is also continuous, where $g : \mathbf{R} \to (-1,1)$ is as defined in Example 5.2. Conversely, if $j : \mathbf{R} \to S$ is continuous, then $j \circ f : (-1,1) \to S$ is continuous, where f is as defined in Example 5.2. Moreover, these two constructions are mutually inverse: Converting h into $h \circ g : \mathbf{R} \to S$ and then to $(h \circ g) \circ f : (-1,1) \to \mathbf{R}$ gives $h \circ (g \circ f) : (-1,1) \to \mathbf{R}$ which is just h, since $g \circ f$ is the identity on $(-1,1)$. Similarly, $(j \circ f) \circ g : \mathbf{R} \to S$ is just j, since $f \circ g$ is the identity on $\mathbf{R}$.

Similarly, a map $k : S \to (-1,1)$ leads to a map $f \circ k : S \to \mathbf{R}$ and so on. $\square$

Lemma 5.4

If $f : S \to T$ is a homeomorphism and $g : T \to U$ is another homeomorphism, then their composite $g \circ f : S \to U$ is also a homeomorphism. Hence if a space T is homeomorphic to S and to U, then S and U must be homeomorphic.

Proof

For $g \circ f$ to be a homeomorphism it must be continuous and have a continuous inverse. Since it is the composite of two continuous maps, it is automatically continuous, by Proposition 3.10.

If f and g are homeomorphisms, then they have continuous inverse maps $f^{-1} : T \to S$ and $g^{-1} : U \to T$. The composite $f^{-1} \circ g^{-1}$ is then continuous, by Proposition 3.10, and inverse to $g \circ f$, since

$$(g \circ f) \circ (f^{-1} \circ g^{-1}) = g \circ (f \circ f^{-1}) \circ g^{-1} = g \circ g^{-1} = id_T$$

and $(f^{-1} \circ g^{-1}) \circ (g \circ f) = id_S$ similarly. Hence $g \circ f$ is a homeomorphism. $\square$

Corollary 5.5

Any open interval of the real line is homeomorphic with $\mathbf{R}$ itself.

Proof

We have seen that $\mathbf{R} \cong (-1, 1)$ and that any two open intervals are homeomorphic. In particular, any open interval is homeomorphic with $(-1, 1)$. Hence, by the lemma, any open interval is homeomorphic with $\mathbf{R}$.   □

Example 5.6

The real projective line $\mathbf{RP}^1$ is homeomorphic with the circle S^1.

We define a function $f : \mathbf{RP}^1 \to S^1$ as follows. A point $l \in \mathbf{RP}^1$ is a straight line through the origin in $\mathbf{R}^2$, and let θ_l be the angle from the positive x-axis to the (first) point where l crosses the circle. (So θ_l is the argument of that point, thinking of points in $\mathbf{R}^2$ as complex numbers.) Then we define $f(l)$ to be the point on the circle whose argument is $2\theta_l$.

To see that f is continuous, let U be an open set in S^1 .We may assume that U is an open arc, since these form a basis for the topology on S^1, so suppose that U is the set of points whose arguments are between ϕ_1 and ϕ_2. Then $f^{-1}(U)$ is the set of lines which cut the circle in points whose arguments are between $\phi_1/2$ and $\phi_2/2$. (Each line cuts the circle in two opposite points; the line is in $f^{-1}(U)$ if *either* of these points has argument between $\phi_1/2$ and $\phi_2/2$.)

By definition of the topology on $\mathbf{RP}^1$ (see Example 3.34), $f^{-1}(U)$ is open if the corresponding subset $\bigcup(l - \{0\})$ is open in $\mathbf{R}^2 - \{0\}$, where the union is taken over all lines $l \in f^{-1}(U)$. This subset consists of two infinite triangles, with their edges removed, placed nose-to-nose:

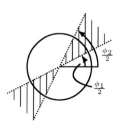

Since the edges are not included in this set, it is clearly open. Hence f is continuous, as the preimage of any basic open set is open.

Thus we have a continuous function $f : \mathbf{RP}^1 \to S^1$. In fact, f is bijective, and an inverse function $g : S^1 \to \mathbf{RP}^1$ can be defined explicitly by setting $g(x, y)$ to be the line which cuts the circle at the point whose argument is half that of (x, y). In other words, $g(\cos\theta, \sin\theta)$ is the line through $(\cos(\frac{\theta}{2}), \sin(\frac{\theta}{2}))$.

The proof that g is continuous is similar to that for f, and is left as Exercise 5.4. Hence f and g are homeomorphisms between $\mathbf{RP}^1$ and S^1.

Example 5.7

If we remove any single point from S^1, then the remaining space is homeomorphic with $\mathbf{R}$. Similarly, if we remove a single point from S^2, the remaining space is homeomorphic with $\mathbf{R}^2$. In either case we can use a technique called **stereographic projection**. We will illustrate this for S^1 with the North Pole removed.

The idea is to draw a copy of $\mathbf{R}$ next to the circle S^1, opposite the omitted point. Then, take straight lines radiating out from that point. These will cross S^1 once and then $\mathbf{R}$, indicating where a point in S^1 is mapped to in $\mathbf{R}$.

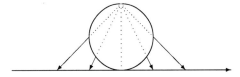

Some elementary geometry shows that this construction takes a point (x, y) in S^1 to $2x/(1 - y)$ in $\mathbf{R}$, thus the corresponding function $S^1 - \{(0, 1)\} \to \mathbf{R}$ is continuous, since it is only being applied to points $(x, y) \in S^1$ where $y \neq 1$. With a little more effort, you can check that a point $x \in \mathbf{R}$ is mapped back to $\left(\frac{4x}{x^2+4}, \frac{x^2-4}{x^2+4}\right)$ in S^1, showing that this inverse operation is also continuous.

This construction can also be used when we remove more than just a single point from S^1 or S^2. For example, if we take the region of S^2 to the south of the equator (or any fixed latitude), then this is homeomorphic to a disc (closed or open according to whether or not we include the equator) under stereographic projection. So, for example, the Southern Hemisphere, including the equator, is homeomorphic to a closed disc in $\mathbf{R}^2$. The part of S^2 south of (and excluding) the Arctic Circle is homeomorphic to an open disc.

If we give $\mathbf{R}$ an extra point, called ∞, then we can extend the homeomorphism $S^1 - \{(0, 1)\} \leftrightarrow \mathbf{R}$ to a bijection $S^1 \leftrightarrow \mathbf{R} \cup \{\infty\}$. We can then put a topology $\mathbf{R} \cup \{\infty\}$ in such a way as to make the extended map a homeomorphism. For example, the open sets containing ∞ would be of the form $(-\infty, -a) \cup (a, \infty) \cup \{\infty\}$, or a union of such a set with an open subset of $\mathbf{R}$.

Similarly, we can construct a homeomorphism between S^2 and $\mathbf{R}^2 \cup \{\infty\}$ (or $\mathbf{C} \cup \{\infty\}$) with a suitable topology. This model of S^2 is called the **Riemann sphere**.

Example 5.8

A solid square is homeomorphic to a solid disc. We will illustrate this with the square

$$Q = \{(x, y) \in \mathbf{R}^2 : -1 \leq x \leq 1, -1 \leq y \leq 1\}$$

and disc

$$D = \{(x, y) \in \mathbf{R}^2 : x^2 + y^2 \leq 1\}.$$

Define $f : D \to Q$ by

$$f(x, y) = \frac{\sqrt{x^2 + y^2}}{\max(|x|, |y|)}(x, y)$$

if $(x, y) \neq (0, 0)$ and $f(0, 0) = (0, 0)$. Its inverse $g : Q \to D$ is given by

$$g(x, y) = \frac{\max(|x|, |y|)}{\sqrt{x^2 + y^2}}(x, y)$$

if $(x, y) \neq (0, 0)$; $g(0, 0) = (0, 0)$.

The idea of these maps is that f pushes the disc out radially to form a square, and g contracts the square radially to form a disc.

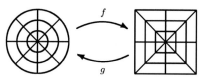

Using this idea you can see that the preimage of an open subset of Q under f will be open in D and similarly for g. So they are continuous maps.

Example 5.9

Similarly, the n-**disc** $D^n = \{(x_1, \ldots, x_n) \in \mathbf{R}^n : \sum_i x_i^2 \leq 1\}$ is homeomorphic with the n-cube $[0, 1]^n = \{(x_1, \ldots, x_n) \in \mathbf{R}^n : 0 \leq x_i \leq 1\}$. Hence, for any space X, the set of continuous maps $D^n \to X$ is in one-to-one correspondence with the set of continuous maps $[0, 1]^n \to X$. We will use this later on to make it easier to define maps on the n-disc.

There are many homeomorphisms like this which are easy to see you when you get the hang of things. The most celebrated is the following.

Example 5.10

A doughnut is homeomorphic to a tea cup. The hole in the doughnut corresponds to the handle on the tea cup. The rest of the tea cup is "homeomorphed away" in the same way that the corners of the square disappeared to form the disc in Example 5.8 - hence the saying that a topologist is someone who cannot tell a doughnut from a tea cup.

But before we get carried away, there are, of course, many pairs of spaces which are not homeomorphic.

Example 5.11

Let $S = \{1, 2\}$ with the discrete topology and let $T = \{1, 2\}$ with the indiscrete topology. If $g : T \to S$ is a bijection, then $g^{-1}\{1\}$ will be a subset containing exactly one point. Any such subset in T is not open, whereas $\{1\} \subset S$ is open. Hence g is not continuous, so there can be no homeomorphism between S and T.

Example 5.12

The closed interval $[0, 1]$ is not homeomorphic with the open interval $(0, 1)$, since Corollary 4.28 says that there is no surjective continuous function $[0, 1] \to (0, 1)$.

Example 5.13

Corollary 4.24 asserts that every continuous map $S^1 \to \mathbf{R}$ is bounded, and so cannot be surjective. Hence S^1 and $\mathbf{R}$ cannot be homeomorphic.

Example 5.14

Corollary 4.19 asserts that every continuous map $\mathbf{R} \to \mathbf{Z}$ is constant, so certainly cannot be bijective. Hence $\mathbf{R}$ and $\mathbf{Z}$ cannot be homeomorphic.

Example 5.15

Theorem 4.1 says that there can be no continuous surjection $\mathbf{R} \to S^0$. Hence $\mathbf{R}$ and S^0 cannot be homeomorphic.

These examples are special cases of a more general phenomenon, which is that homeomorphic spaces share all topological properties. Hence if two spaces do not share a topological property, then they cannot be homeomorphic.

Proposition 5.16

Let S and T be two spaces which are homeomorphic. Then

1. If S is connected, then so is T.

2. If S is compact, then so is T.

3. If S is Hausdorff, then so is T.

Proof

1. The quickest way to prove this is to use Proposition 4.11, which says that there can be no continuous surjection from a connected space to a disconnected space. If S and T are homeomorphic, then there certainly is a continuous surjection $S \to T$ (a bijection, in fact), so T cannot be disconnected.

2. Suppose that $f : S \to T$ is a homeomorphism. By Proposition 4.27, the image of f is compact. If f is a homeomorphism, then this image is T. Hence T is compact.

3. Again, suppose that $f : S \to T$ is a homeomorphism. The inverse homeomorphism $T \to S$ is continuous and injective. As S is Hausdorff, it follows from Proposition 4.35 that T is Hausdorff.   □

This gives us numerous examples of non-homeomorphic pairs of spaces.

Example 5.17

The 0-sphere S^0 and the circle S^1 are not homeomorphic, as S^1 is connected and S^0 is disconnected.

Example 5.18

The closed interval $[0, 1]$ is not homeomorphic to the real line $\mathbf{R}$ since $[0, 1]$ is compact and $\mathbf{R}$ is not.

Example 5.19

The real line $\mathbf{R}$ is not homeomorphic to the real line with a double point (Example 4.37) since $\mathbf{R}$ is Hausdorff, but the real line with a double point is not.

We can also adapt this approach to handle spaces that share a property, if a simple modification of the spaces yields two spaces which do not share that property.

Example 5.20

The circle S^1 is not homeomorphic to the closed interval $[0, 1]$.

This is because if we remove the point $1/2$ from $[0, 1]$ then we end up with a disconnected space, whereas if we remove any point from S^1, then we still have a connected space. If they were homeomorphic before removing a point from each, then they will be homeomorphic afterwards, but connectivity tells us that this cannot be so.

To prove this rigorously, assume that $f : S^1 \to [0, 1]$ is a homeomorphism, with $g : [0, 1] \to S^1$ its inverse.

Let $P = [0, 1/2) \cup (1/2, 1]$ and let $T = S^1 - g(1/2)$, so the image of P under g is T and the image of T under f is P. Therefore, if P and T have the subspace topologies, the maps f and g restrict to continuous bijections $P \longleftrightarrow T$. However, P is disconnected: set $U = [0, 1/2)$ and $V = (1/2, 1]$. But T is connected as it is homeomorphic to $(0, 1)$ (by stereographic projection). Therefore, P and T cannot be homeomorphic, so S^1 and $[0, 1]$ cannot have been homeomorphic in the first place.

Example 5.21

Adapting Example 5.20, we can see that $[0, 1)$ is not homeomorphic with S^1 for, if it were, then deleting a single point from each would leave homeomorphic spaces. But if we delete the point $1/2$ from $[0, 1)$, we get a disconnected space, whereas deleting any single point from S^1 will leave a connected space. So $[0, 1)$ and S^1 cannot be homeomorphic.

A pitfall that we need to be aware of is that a continuous function may be bijective but not a homeomorphism, as its inverse may fail to be continuous.

Example 5.22

Let $e : [0, 1) \rightarrow S^1$ be the restriction of the exponential map of Example 3.47

$$e(t) = (\cos(2\pi t), \sin(2\pi t)).$$

As the restriction of a continuous map, this is continuous, and it is clear that e is a bijection. However, e is *not* a homeomorphism because, as we have just seen, $[0, 1)$ and S^1 are *not* homeomorphic.

We can see that e is not a homeomorphism directly, in the following way. Since e is bijective, there is no choice in the definition of its inverse map, e^{-1} must send $(\cos(\theta), \sin(\theta))$ to $\theta/2\pi \in [0, 1)$. However, the half-open interval $[0, 1/2)$ is an *open* subset of $[0, 1)$ under the subspace topology, and its preimage under this inverse map e^{-1} will be the set of points with positive y-coordinate, together with the point $(1, 0)$. This is *not* an open subset of S^1, since any open ball around $(1, 0)$ would include some points with negative y-coordinate. So the map e^{-1} is not continuous, and thus e is not a homeomorphism.

In fact, even between homeomorphic spaces, it is possible to construct a continuous bijection that is not a homeomorphism, as the following example shows:

Example 5.23

Let $V = (0, 1] \cup (2, 3] \cup (4, 5] \cup \ldots$, and let $f : V \rightarrow V$ be the function defined by

$$f(x) = \begin{cases} \frac{x}{2} & \text{if } x \in (0, 1], \\ \frac{x-1}{2} & \text{if } x \in (2, 3], \\ x - 2 & \text{otherwise.} \end{cases}$$

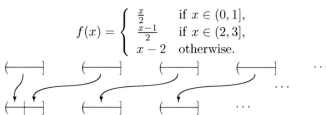

A quick check shows that f is continuous, and bijective. But its inverse would have to map the interval $(0, 1]$ to $(0, 1] \cup (2, 3]$, i.e., a connected space to a disconnected space. This cannot happen if the inverse is continuous.

The problem in both of these examples is that continuity ensures that the preimage of every open set is open, but not that the image of every open set is open. Putting that another way, continuity ensures that the function gives a surjection from the collection of open sets of the domain to the collection of open sets of the range, but a homeomorphism must give a *bijection* of open sets, not just a surjection. A map $f : S \rightarrow T$ with the property that $f(U)$ is open in T whenever U is an open set in S is called an **open map**. Thus a continuous bijection which is an open map is a homeomorphism. The following theorem shows that, under certain conditions, all continuous bijections are open maps.

Theorem 5.24

If X is a compact space, Y is a Hausdorff space and $f : X \to Y$ is a continuous function which is bijective, then there is a continuous inverse function $g : Y \to X$ with $gf = 1_X$ and $fg = 1_Y$. Hence f is a homeomorphism.

We will prove this using three lemmas.

Lemma 5.25

If X is compact and $U \subset X$ is a closed subspace, then U is compact.

Proof

Suppose we take any open cover of U. As U is a subspace, each set in this cover is the intersection of U with an open set of X. These open subsets of X need not form a cover of X, but if we include the complement $X - U$ (which is open, since U is closed), then we obtain an open cover of X. Because X is compact, we can refine this cover to a finite open cover. If we omit $X - U$, and intersect each set in this cover with U, then we obtain a finite refinement of the original open cover of U. Hence U is compact.   □

Lemma 5.26

If Y is Hausdorff and $V \subset Y$ is a compact subspace, then V is closed.

Proof

We will show that the complement $Y - V$ is open by constructing, for each $y \in Y - V$, an open set of $Y - V$ containing y. To do this, let $v \in V$ be any point. Since $y \in Y - V$, so y and v must be distinct points. As Y is Hausdorff, we can therefore find disjoint open sets $U_{y,v}$ and U_v of Y such that $y \in U_{y,v}$ and $v \in U_v$.

By doing this for all $v \in V$, we obtain a bunch of open sets $U_v \subset Y$ such that every $v \in V$ belongs to at least one, namely U_v. If we intersect each U_v with V, then we obtain an open cover of V. Since V is compact, we can refine this cover to a finite list of open sets, say $U_{v_1}, U_{v_2}, \ldots, U_{v_n}$ for some n. Now, we have some (possibly different) open sets $U_{y,v_1}, U_{y,v_2}, \ldots, U_{y,v_n}$ of Y, each of which contains y. Since there are only finitely many, we can intersect them to get an open set U_y of Y which still contains y. Now U_y is disjoint from each U_{v_i} and, since these U_{v_i} cover V, U_y is disjoint from V. In other words, $U_y \subset Y - V$. Hence we have an open subset of Y containing y and contained in $Y - V$. Since we can do this for each $y \in Y$, we see that $Y - V$ is open, i.e., V is closed. $\square$

Lemma 5.27

Let $f : S \to T$ be a function between two topological spaces. Then f is continuous if, and only if, $f^{-1}(U)$ is closed whenever U is a closed subset of T.

Proof

First let us assume that f is continuous, i.e., $f^{-1}(V)$ is open whenever V is an open subset of T. If $U \subset T$ is closed, then $T - U$ is open, so $f^{-1}(T - U)$ is open. However, $f^{-1}(T - U) = S - f^{-1}(U)$, so the complement of $f^{-1}(U)$ is open, i.e., $f^{-1}(U)$ is closed.

By the same argument, if $f^{-1}(U)$ is closed whenever U is closed, then the preimage of any open set is open, i.e., f is continuous. $\square$

Proof (of Theorem 5.24)

Let $f : X \to Y$ be a continuous bijection. Since it is bijective, its inverse map $g : Y \to X$ is already given, and all we have to do is show that g is continuous. We will do this using Lemma 5.27, so let $U \subset X$ be a closed set. Since X is compact, Lemma 5.25 shows that U is compact. Now the preimage $g^{-1}(U)$ is exactly $f(U)$ which, by Proposition 4.27, is compact, being the image of a compact set under a continuous map. Thus $f(U)$ is a compact subset of a Hausdorff space Y hence, by Lemma 5.26, $g^{-1}(U) = f(U)$ is closed. Thus the preimage of the closed set U is closed. $\square$

5.2 Disjoint Unions

Having established what we mean by saying two topological spaces are "the same", we now turn to ways of building new topological spaces out of old ones which, we hope, will enable us to construct the space we're interested in, or something homeomorphic to it.

The simplest topological construction is that of forming "disjoint unions", where we take two topological spaces and, while keeping them separate, think of them as a single space.

Example 5.28

The space $S^0 = \{-1, +1\}$ is the disjoint union of two one-point spaces $\{-1\}$ and $\{+1\}$:

$$S^0 = \{-1\} \amalg \{+1\}.$$

If we have any two topological spaces S, T, then we can define their **disjoint union** $S \amalg T$ as follows. The points in $S \amalg T$ are given by taking all the points of S together with all the points in T, and thinking of all these points as being distinct. So if the sets S and T overlap, then each point in the intersection $S \cup T$ occurs twice in the disjoint union $S \amalg T$. We can therefore think of S as a subset of $S \amalg T$ and we can think of T as a subset of $S \amalg T$, and these two subsets do not intersect. We then topologize $S \amalg T$ by defining a subset Q of $S \amalg T$ to be open if $Q \cap S$ is an open subset of S and $Q \cap T$ is an open subset of T. So the open sets of $S \amalg T$ are just the unions of an open set in S with an open set in T.

Example 5.29

We can form a set of train tracks by taking $\mathbf{R} \amalg \mathbf{R}$:

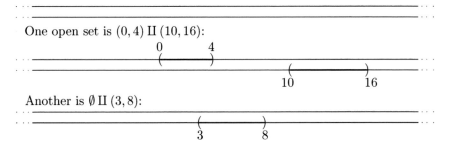

One open set is $(0, 4) \amalg (10, 16)$:

Another is $\emptyset \amalg (3, 8)$:

Note that if S and T are both subsets of a given space and $S \cap T$ is not empty, then in $S \amalg T$ we count the points in the intersection *twice*. This is illustrated by the following example.

Example 5.30

The disjoint union of the circle S^1 and the interval $[-2, 2]$ is homeomorphic to a "rising sun":

and not homeomorphic to the *union* of S^1 with $[-2, 2] \times \{0\}$.

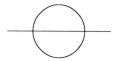

The disjoint union is quite a simple construction as it is quite easy to describe continuous maps to or from disjoint unions.

Theorem 5.31

If R is any topological space, then a continuous map $S \amalg T \to R$ corresponds to a pair of continuous maps $(S \to R, T \to R)$.

If Q is a connected topological space, then a continuous map $Q \to S \amalg T$ corresponds to either a continuous map $Q \to S$ or a continuous map $Q \to T$.

Proof

If $g : S \amalg T \to R$, then we can define continuous maps $g_S : S \to R$ and $g_T : T \to R$ by

$$g_S(x) = g(x) \text{ for } x \in S \quad \text{and} \quad g_T(y) = g(y) \text{ for } y \in T.$$

On the other hand, if $h_S : S \to R$ and $h_T : T \to R$ are continuous maps, then we can define a continuous map $h : S \amalg T \to R$ by

$$h(x) = \begin{cases} h_S(x) & \text{if} \quad x \in S, \\ h_T(x) & \text{if} \quad x \in T. \end{cases}$$

If $f : Q \to S \amalg T$ is continuous, then the preimages $f^{-1}(S)$ and $f^{-1}(T)$ will both be open. Their union is $f^{-1}(S \amalg T) = Q$, and their intersection is empty. So, as Q is connected, either $f^{-1}(S)$ or $f^{-1}(T)$ must be empty. If $f^{-1}(S) = \emptyset$, then the image of f is contained in T, so we can think of f as a continuous map $Q \to T$. If, on the other hand, $f^{-1}(T) = \emptyset$, then f corresponds to a continuous map $Q \to S$. $\qquad\square$

Example 5.32

The space S^0 is a disjoint union $S^0 = \{-1\} \amalg \{+1\}$, so a continuous map $S^0 \to \mathbf{R}$, for example, corresponds to a pair of continuous maps $\{-1\} \to \mathbf{R}$ and $\{+1\} \to \mathbf{R}$. A continuous map from a one-point space to $\mathbf{R}$ is determined by its image, which will be a point in $\mathbf{R}$. And all points in $\mathbf{R}$ correspond to continuous maps in this way. Thus continuous maps $S^0 \to \mathbf{R}$ correspond to ordered pairs of real numbers, i.e., points in $\mathbf{R}^2$.

Example 5.33

The space $\mathbf{R} - \{0\}$ of non-zero real numbers is homeomorphic to a disjoint union of two copies of $\mathbf{R}$, i.e.,

$$\mathbf{R} - \{0\} \cong \mathbf{R} \amalg \mathbf{R}.$$

We can define a function $f : \mathbf{R} \amalg \mathbf{R} \to \mathbf{R} - \{0\}$ using Theorem 5.31 as follows. Let $f_1 : \mathbf{R} \to (0, \infty)$ be a homeomorphism, e.g., $f_1(x) = e^x$, and let $f_2 : \mathbf{R} \to (-\infty, 0)$ be another homeomorphism, e.g., $f_2(x) = -e^x$. Then define f to equal f_1 on the first copy of $\mathbf{R}$ and f_2 on the second copy of $\mathbf{R}$. By the theorem, this is continuous, and it is obviously bijective as well.

The inverse map $g : \mathbf{R} - \{0\} \to \mathbf{R} \amalg \mathbf{R}$ is also continuous. For if Q is an open set of $\mathbf{R} \amalg \mathbf{R}$, then $Q = S \amalg T$ where S and T are open sets in $\mathbf{R}$. The preimage $g^{-1}(Q)$ is then $g_1^{-1}(S) \cup g_2^{-1}(T)$, where g_1 and g_2 are the inverses of f_1 and f_2 respectively. Since f_1 and f_2 are homeomorphisms, so g_1 and g_2 are continuous, hence these preimages are open and, consequently, so is their union. Thus g is continuous, and f and g are homeomorphisms.

Example 5.34

The group $O(3)$ of orthogonal matrices is homeomorphic with a disjoint union of two copies of the group $SO(3)$ of special orthogonal matrices. One copy is the natural subset of $O(3)$ consisting of matrices with determinant $+1$. By definition this is $SO(3)$. On the other hand, the set of matrices with determinant -1 in $O(3)$ is homeomorphic to $SO(3)$, by the relation $M \mapsto -M$. As M is a 3×3 matrix, $\det(-M) = (-1)^3 \det(M) = -\det(M)$ and if M is orthogonal, then so is $-M$. And, clearly, the two copies of $SO(3)$ are disjoint.

The relation between the disjoint union construction and compactness is quite simple.

Theorem 5.35

If S and T are compact topological spaces, then their disjoint union $S \amalg T$ will be compact. Conversely, if $S \amalg T$ is compact, then both S and T must be compact.

Proof

To prove the first assertion, take any open cover $\mathcal{U}$ of $S \amalg T$. By definition of the topology on $S \amalg T$, each open set in $\mathcal{U}$ will be a union of an open subset of S and an open subset of T. So we have, in particular, an open cover of S and an open cover of T. As both S and T are compact, we can finitely refine each of these open covers. Let $\mathcal{V}$ be the collection of all sets Q in $\mathcal{U}$ such that either $Q \cap S$ is part of the finite refinement of the open cover of S, or, $Q \cap T$ is part of the finite refinement of the open cover of T.

Then $\mathcal{V}$ will be an open cover of $S \amalg T$ as it covers both S and T. Moreover, it will be finite, as the most sets it could include would involve one for each set in the finite refinement of the cover of S and one for each set in the finite refinement of the cover of T, i.e., the sum of two finite numbers which will again be finite.

Conversely, suppose that $S \amalg T$ is compact. Let $\mathcal{U}$ be an open cover of S. We can turn this into an open cover for $S \amalg T$ by also including the set T. Since $S \amalg T$ is compact, this open cover for $S \amalg T$ has a finite refinement. This refinement will include T, but if we remove T from it, we will be left with a finite refinement of the original open cover for S. Hence S is compact, and the same argument can be applied to T. □

The following result about the Hausdorff property is proved in a similar way.

Proposition 5.36

If S and T are Hausdorff topological spaces, then their disjoint union $S \amalg T$ will be Hausdorff. Conversely, if $S \amalg T$ is Hausdorff, then both S and T must be Hausdorff.

Connectivity is, obviously, rather different: A disjoint union is never connected. In fact, the two notions are intimately related, because a disconnected space can always be expressed as a disjoint union of two non-trivial spaces.

Lemma 5.37

If S is disconnected, then there are non-empty spaces Q, R such that $S \cong Q \amalg R$. Conversely, if Q and R are non-empty, then their disjoint union $Q \amalg R$ is disconnected.

Proof

If S is disconnected, then there are non-empty open sets U, V contained in S such that $U \cup V = S$ and $U \cap V = \emptyset$. Let Q be the subset U with the subspace topology, and V the subset R with the subspace topology. Then, by definition of the topology on a disjoint union, S is homeomorphic with $Q \amalg R$.

Conversely, in $Q \amalg R$, let $U = Q$ and $V = R$. By hypothesis these are non-empty, and by the definition of the disjoint union they are disjoint and open. $\qquad\square$

Thus, if a space is not connected, we can decompose it into two separate components. It is tempting to think that we can continue this process, and express any space as a disjoint union of connected subspaces. However, this is not the case, because this splitting process may never end, as the space $\mathbf{Q}$ shows. The only connected subspaces of $\mathbf{Q}$ are those containing a single point. However, no such subspace is open, as $\mathbf{Q}$ is not discrete. Thus, as a topological space, $\mathbf{Q}$ cannot be expressed as the disjoint union of its connected subspaces.

5.3 Product Spaces

A slightly more intricate way of combining two topological spaces is the "product construction". If we have two sets S, T, then we can form their **Cartesian product** $S \times T$ whose points are pairs (s, t), with one point s from S and one point t from T.

Example 5.38

The most familiar example of this is $\mathbf{R} \times \mathbf{R}$, which we usually write as $\mathbf{R}^2$. A point in $\mathbf{R}^2$ is, by definition, a pair (x, y) of one number, x, taken from $\mathbf{R}$, and another number y, also taken from $\mathbf{R}$. This is exactly how we have just described the product $\mathbf{R} \times \mathbf{R}$.

But if we want to take the product of two topological spaces, and we want the result to be a topological space again, then we need to put a topology on $S \times T$, based on the topologies of S and T. And, of course, we want this topology to be such that when we take $\mathbf{R} \times \mathbf{R}$ we get the familiar topology on $\mathbf{R}^2$. We will use this example to guide us in putting a topology on any product $S \times T$.

Recall that a basis for $\mathbf{R}^2$ is given by taking all the open discs in $\mathbf{R}^2$. We will first indicate how to construct an open disc from the topology on $\mathbf{R}$. The obvious open sets in $\mathbf{R}^2$ are those of the form $P \times Q$ where P and Q are open sets of $\mathbf{R}$. In particular, if P and Q are both open intervals, then $P \times Q$ is an open rectangle in $\mathbf{R}^2$. We can use these to construct an open disc in $\mathbf{R}^2$ by taking the union of all the open rectangles whose corners lie on the circumference of the disc.

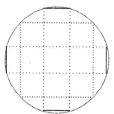

An open disc as a union of open rectangles

Thus every open disc can be expressed as a union of sets of the form $P \times Q$ where P and Q are open sets. Consequently *every* open set of $\mathbf{R}^2$ can be expressed as such a union. In other words, the sets of the form

$$(\text{open set}) \times (\text{open set})$$

give a basis for the topology on $\mathbf{R}^2$.

In general, then, if we have any two topological space S and T, the **product topology** on $S \times T$ is the topology which has a basis consisting of all products $P \times Q$ of an open subset $P \subset S$ with an open subset $Q \subset T$.

Example 5.39

The discussion above shows that the topology on $\mathbf{R} \times \mathbf{R}$ is the same as that on $\mathbf{R}^2$.

Example 5.40

Similarly, $\mathbf{R}^m \times \mathbf{R}^n$ is homeomorphic with $\mathbf{R}^{m+n}$ by the correspondence $((x_1, \ldots, x_m), (y_1, \ldots, y_n)) \leftrightarrow (x_1, \ldots, x_m, y_1, \ldots, y_n)$.

Example 5.41

The cylinder C of Example 3.28 is homeomorphic to the product $S^1 \times [0,1]$.

To construct maps between C and $S^1 \times [0,1]$, we use the fact that $\mathbf{R}^2 \times \mathbf{R}$ is homeomorphic to $\mathbf{R}^3$ by maps $((x,y),z) \leftrightarrow (x,y,z)$. Restricting these to $C \subset \mathbf{R}^3$ and to $S^1 \times [0,1] \subset \mathbf{R}^2 \times \mathbf{R}$ gives bijective maps $C \leftrightarrow S^1 \times [0,1]$ as required.

In order to recognize other spaces as homeomorphic to product spaces, it will be useful to have the following simple way of describing maps to product spaces.

Theorem 5.42

A continuous map $f : Q \to S \times T$ corresponds to a pair of continuous maps $f_1 : Q \to S$ and $f_2 : Q \to T$.

Proof

Given $f : Q \to S \times T$, we define f_1 and f_2 to be the composites $f_1 = p_1 \circ f$ and $f_2 = p_2 \circ f$, where $p_1 : S \times T \to S$ and $p_2 : S \times T \to T$ are the projections defined by $p_1(s,t) = s$ and $p_2(s,t) = t$. To see that these are continuous, note that if $U \subset S$ is an open set, then $p_1^{-1}(U) = U \times T$ which is an open set in the product topology (it is even in the basis for this topology), and similarly for p_2. Thus f_1, f_2 are composites of continuous maps and, hence, continuous.

Conversely, if we are given f_1, f_2, then define f by $f(q) = (f_1(q), f_2(q))$. To see that this is continuous, we use Proposition 3.44 so that we only need to check that the preimages of basic open sets are open. So, let $U \times V$ be a basic open set, i.e., $U \subset S$ is open and $V \subset T$ is open, then $f^{-1}(U \times V) = f_1^{-1}(U) \cap f_2^{-1}(V)$, the intersection of two open sets if f_1 and f_2 are continuous and, hence, open.

Since these two operations of splitting f into a pair (f_1, f_2) and of turning a pair (f_1, f_2) into a combined function f are mutually inverse, so we get the one-to-one correspondence of the Theorem.   □

Example 5.43

The torus T^2 is homeomorphic to the product $S^1 \times S^1$.

For example, a map $f : S^1 \times S^1 \to T^2$ is given by

$$f((x,y),(x',y')) = ((x'+2)x, (x'+2)y, y').$$

This is continuous, being given by additions and multiplications, and its inverse map $g : T^2 \to S^1 \times S^1$ is given by

$$g(x,y,z) = \left(\left(\frac{x}{\sqrt{x^2+y^2}}, \frac{y}{\sqrt{x^2+y^2}} \right), (\sqrt{x^2+y^2} - 2, z) \right).$$

This can be seen to be continuous if you trust that the square-root function is continuous. Alternatively, you can use Theorem 5.24 since T^2 is compact, and Theorem 5.47 shows that $S^1 \times S^1$ is Hausdorff.

Example 5.44

The space $\mathbf{R}^2 - \{0\}$ is homeomorphic to the product $S^1 \times (0, \infty)$. For we can define maps $f : \mathbf{R}^2 - \{0\} \to S^1 \times (0, \infty)$, $g : S^1 \times (0, \infty) \to \mathbf{R}^2 - \{0\}$ by

$$f(x,y) = \left(\frac{(x,y)}{\sqrt{x^2+y^2}}, \sqrt{x^2+y^2} \right) \qquad \text{and} \qquad g((x,y),t) = (tx,ty).$$

Example 5.45

The **annulus**

$$A = \{(x,y) \in \mathbf{R}^2 : 1 \le \sqrt{x^2+y^2} \le 2\}$$

is homeomorphic to the product $S^1 \times [1,2]$ by the same pair of maps.

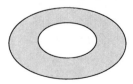

Example 5.46

The train-track space $\mathbf{R} \amalg \mathbf{R}$ of Example 5.29 is homeomorphic to $\mathbf{R} \times S^0$.

The relation between the product topology and the topological properties that we have met is very simple.

Theorem 5.47

If S and T are topological spaces, then

1. The product $S \times T$ is Hausdorff if, and only if, S and T are both Hausdorff.

2. The product $S \times T$ is connected if, and only if, S and T are both connected.

3. The product $S \times T$ is compact if, and only if, S and T are both compact.

Proof

For each part one of the implications is hard and the other is easy. We will prove the hard ones, leaving the easy implications as Exercise 5.9.

First we will show that $S \times T$ is Hausdorff only if S and T are both Hausdorff. So let us assume that $S \times T$ is Hausdorff. To show that S is Hausdorff, let s, s' be two distinct points in S, and let t be any point in T. Thus $(s,t) \neq (s',t)$, so there are non-overlapping open subsets Q, $Q' \subset S \times T$ such that $(s,t) \in Q$ and $(s',t) \in Q'$. As Q is open in the product topology, it must be a union of subsets of the form $U \times V$ where U is an open subset of S and V is an open subset of T. In particular, (s,t) must lie in such a subset, so that $s \in U$. Similarly, (s',t) must lie in some subset $U' \times V'$ where U' is an open subset of S. Now, if $U \cap U'$ is non-empty, then there is some element $s'' \in U \cap U'$, in which case $(s'',t) \in U \cap V$ since $t \in V$, and $(s'',t) \in U' \cap V'$ since $t \in V'$. Hence $U \cap V$ and $U' \cap V'$ overlap. This cannot happen, so the intersection $U \cap U'$ must, in fact, be empty. Hence S is Hausdorff since we have found, for any pair of distinct points $s, s' \in S$, a non-overlapping pair of open sets U, U' each containing only one of the points. Similarly, T is Hausdorff.

Next we will show that $S \times T$ is connected if S and T are both connected. To do this, suppose that $f : S \times T \to S^0$ is continuous. Let $s_0 \in S$ be any point; then the subspace $\{s_0\} \times T \subset S \times T$ is homeomorphic to T and, hence, connected. So the restriction of f to $\{s_0\} \times T$ must be constant; let's assume that $f(s_0, t) = 1$ for all $t \in T$. Now, for each $t \in T$, we can restrict f to the subset $S \times \{t\}$. Each such subset is, as a subspace of $S \times T$, homeomorphic with S and consequently connected. So f is constant on each such subset. But $(s_0, t) \in S \times \{t\}$, and $f(s_0, t) = 1$. Hence $f(s,t) = 1$ for all $(s,t) \in S \times \{t\}$. The same argument applies for all $t \in T$; hence $f(s,t) = 1$ for all $(s,t) \in S \times T$. In other words, f is constant on $S \times T$. Hence there can be no continuous surjection $S \times T \to S^0$, so $S \times T$ is connected.

Finally, we will show that $S \times T$ is compact if S and T are both compact. This result is known as **Tychonov's theorem** and is proved using a few tricks. To show that $S \times T$ is compact, let us assume that we have an open cover $\{W_i\}_{i \in I}$ of $S \times T$, where I is some indexing set, presumably infinite. Thus every point (s,t) in $S \times T$ lies in one of the open sets W_i. Since W_i is open in the product topology, there is some set $S_{s,t} \times T_{s,t}$, where $S_{s,t}$ is an open set of S and $T_{s,t}$ is an open set of T, such that $(s,t) \in S_{s,t} \times T_{s,t} \subset W_i$. By taking all points $(s,t) \in S \times T$, the sets $\{S_{s,t} \times T_{s,t}\}_{(s,t) \in S \times T}$ form an open cover for $S \times T$. If we can refine this cover, then we can obtain a corresponding refinement of the cover $\{W_i\}_{i \in I}$ since each set $S_{s,t} \times T_{s,t}$ is contained in one of the sets W_i. Consequently, we can assume that our open cover is of the form $\{S_i \times T_i\}_{i \in I}$, i.e., each set in the cover is the product of an open set in S with an open set in T.

Now, for each $s \in S$, the subset $\{s\} \times T$ has an open cover given by intersecting $\{s\} \times T$ with the open cover $\{S_i \times T_i\}_{i \in I}$ of $S \times T$. Since $\{s\} \times T$ is homeomorphic with T, it is compact, so there is some finite refinement $S_{i_1} \times T_{i_1}$, ..., $S_{i_n} \times T_{i_n}$ which, when intersected with $\{s\} \times T$, gives a cover of this subspace. Now let $S^s = S_{i_1} \cap S_{i_2} \cap \cdots \cap S_{i_n}$. This is a finite intersection of open sets of S, so is again an open set of S, and it contains s. By carrying out this process for each $s \in S$, we thus obtain an open cover $\{S^s\}_{s \in S}$ of S. Since S is compact, this cover can be refined, so we have a finite list of points $s_1, \ldots, s_m$ of S such that $S^{s_1}, \ldots, S^{s_m}$ cover S. For each of these points, the process above gave a finite refinement of the cover of $S \times T$ which covers $S^{s_i} \times T$. Thus, if we take all of these finite refinements, for each of the (finitely many) points $s_1, \ldots, s_m$, we will obtain a collection of open sets which cover $S \times T$ and which are a finite refinement of the original cover $\{S_i \times T_i\}_{i \in I}$. Hence $S \times T$ is compact, as required. □

5.4 Quotient Spaces .

The final construction that we will cover in this chapter is that of forming "quotient spaces". This construction is very useful, but is harder to understand than disjoint unions and products, so we will begin by looking in detail at an example.

Example 5.48

Suppose we take the closed disc D^2 in the plane $\mathbf{R}^2$,

$$D^2 = \{(x, y) \in \mathbf{R}^2 : \sqrt{x^2 + y^2} \le 1\}.$$

Its boundary, which is traditionally written ∂D^2, is exactly the space S^1. The "quotient space" $D^2/\partial D^2$ is the space obtained by considering all the points on the boundary as equal. It is as if there were a drawcord around the perimeter of D^2 which we pull tight.

This suggests that the resulting space should be homeomorphic to the 2-sphere S^2, but how do we put a topology on the quotient space so that this is the case?

Before we can put a topology on this "quotient space" we'd best be completely clear about what the set is that we are dealing with. Concretely, then, let $D^2/\partial D^2$ denote the set $D^2 - \partial D^2$ together with a single extra point $\{*\}$. So we remove the boundary from D^2, and replace it by a single point.

With this as our set, we can describe the process of pulling the drawcord tight as a function $\pi : D^2 \to D^2/\partial D^2$ defined as follows. We send every point $x \in D^2 - \partial D^2$ to its corresponding point in $D^2/\partial D^2$, and we send every point of the boundary $\partial D^2 \subset D^2$ to the extra point $*$.

Now, as $D^2/\partial D^2$ is, so far, just a set, and not a topological space, it makes no sense to ask whether or not this function π is continuous. However, we would *like* it to be continuous, so that as the boundary of D^2 maps to the point $*$, so all points near the boundary map to points near $*$.

So we should put a topology on $D^2/\partial D^2$ which makes π continuous. Well, we know one topology which we can put on $D^2/\partial D^2$ that will make this function continuous: The indiscrete topology. According to Proposition 3.9, any function whose range has the indiscrete topology is continuous. But this is too brutal as it completely ignores the topology that we have on D^2. Instead we should give $D^2/\partial D^2$ as many open sets as possible subject to the condition that the map from D^2 be continuous. In other words, a subset $U \subset D^2/\partial D^2$ is open if its preimage $\pi^{-1}(U)$ is open in D^2.

This gives two sorts of open sets in $D^2/\partial D^2$ – those which contain the point $*$ and those which do not. A subset $S \subset D^2/\partial D^2$ which does not contain $*$ can be thought of as a subset of $D^2 - \partial D^2$, and S will be open precisely if the corresponding subset of $D^2 - \partial D^2$ is open. On the other hand, if $S \subset D^2/\partial D^2$ does contain $*$, then its complement, $D^2/\partial D^2 - S$, does not contain $*$, and so we can think of this complement as a subset of $D^2 - \partial D^2$ and, consequently, as a subset of D^2. Then S will be open precisely if this complement is closed when considered as a subset of D^2.

Phrasing this in general terms then, if X is a topological space and A is a subset of X, then the **quotient space** X/A is the set $(X - A) \amalg \{*\}$, and a subset $U \subset (X - A) \amalg \{*\}$ is open if, and only if, $\pi^{-1}(U)$ is open in X, where $\pi : X \to (X - A) \amalg \{*\}$ is the function defined by $\pi(x) = x$ if $x \notin A$ and $\pi(x) = *$ if $x \in A$. Hence the open sets in X/A are either open sets in $X - A$, or unions of $\{*\}$ and the intersection with $X - A$ of an open set in X containing A.

Example 5.49

Let's check that, with this topology, $D^2/\partial D^2$ is homeomorphic with S^2 as we claimed. First, note that $D^2 - \partial D^2$ is homeomorphic with $\mathbf{R}^2$, by

$$(r\cos(\theta), r\sin(\theta)) \longleftrightarrow \left(\tan(\frac{\pi r}{2})\cos(\theta), \tan(\frac{\pi r}{2})\sin(\theta)\right)$$

(compare with Example 5.2). And $\mathbf{R}^2$ is homeomorphic to $S^2 - \{(0,0,1)\}$ by stereographic projection. By composing these, we can get a homeomorphism $f : D^2 - \partial D^2 \cong S^2 - \{(0,0,1)\}$. We can extend this to a bijection $g : D^2/\partial D^2 \cong S^2$ by $* \leftrightarrow (0,0,1)$. However, we need to show that g and its inverse are continuous.

If $Q \subset S^2$ is an open set not containing $(0,0,1)$, then $g^{-1}(Q) = f^{-1}(Q)$ is an open set in $D^2 - \partial D^2$, hence $g^{-1}(Q)$ is an open set in $D^2/\partial D^2$. On the other hand, if $Q \subset S^2$ contains $(0,0,1)$, then $g^{-1}(Q) = \{*\} \cup f^{-1}(Q')$, where $Q' = Q - \{(0,0,1)\}$. Since Q is open, it must contain some breathing space around $(0,0,1)$. Under the homeomorphism $S^2 - \{(0,0,1)\} \leftrightarrow \mathbf{R}^2$, this breathing space (excluding $(0,0,1)$) corresponds to the set of points *outside* a certain disc in $\mathbf{R}^2$, i.e., all points sufficiently far from the origin. And $Q - \{(0,0,1)\}$ then corresponds to the complement of some closed and bounded subset of $\mathbf{R}^2$. Hence, $f^{-1}(Q - \{(0,0,1)\})$ in $D^2 - \partial D^2$ is the complement of some closed subset contained within some disc $B_\delta(0,0)$ with $\delta < 1$, i.e., not touching the boundary. Hence, if we take the union of this with ∂D^2, we get an open set in D^2. In other words, $g^{-1}(Q)$ is the union of $\{*\}$ with the intersection to $D^2 - \partial D^2$ of an open set containing ∂D^2. Hence $g^{-1}(Q)$ is open even in this case, and thus g is continuous.

Similarly, the inverse g^{-1} is also continuous.

Sometimes squares are easier to deal with than discs, and the following two examples will be very useful later on.

Example 5.50

If X is the product $X = [0,1]^2 = [0,1] \times [0,1]$ of two copies of the closed interval $[0,1]$, and $A = \partial X = \{(x,y) \in [0,1]^2 : x(1-x)y(1-y) = 0\}$ is its boundary, then X/A is homeomorphic to S^2. This can be shown using a homeomorphism between $[0,1]^2$ and D^2, such as given by Example 5.8, which takes A to the boundary ∂D^2 of D^2. Hence $X/A \cong D^2/\partial D^2 = S^2$.

Example 5.51

More generally, let n be any positive integer, and let X be the product of n copies of the interval $[0,1]$, i.e., $X = [0,1]^n$, and let $A = \partial X$ be the boundary of this, so that

$$A = \{(x_1, \ldots, x_n) \in [0,1]^n : x_i = 0 \text{ or } x_i = 1 \text{ for at least one } i\}.$$

Then X/A is homeomorphic with the n-sphere S^n.

Example 5.52

The quotient space $\mathbf{R}/\mathbf{Z}$ is a "bouquet" of infinitely many circles (one for each integer) all joined together at one point.

Note that the quotient space does not take account of any algebraic structure there may be: $\mathbf{R}$ and $\mathbf{Z}$ are both groups, and their quotient group $\mathbf{R}/\mathbf{Z}$ is isomorphic to the group of complex numbers of length 1, i.e., S^1. So the quotient *space* $\mathbf{R}/\mathbf{Z}$ is quite different to the quotient *group* $\mathbf{R}/\mathbf{Z}$.

A generalization of this sort of quotient construction is where we glue some points together, but not necessarily all to the same point. For example, if we take a square then we should be able to get a cylinder by gluing all the points along one edge to the corresponding points on the other edge. While this doesn't fit directly into the quotient construction framework outlined above, it can be dealt with in a very similar way.

We first have to say which points are glued to which. If two points a and b are to be glued together, and b and c are also to be glued together, then clearly a and c will be glued together, too. So we should have an equivalence relation on our space.

Example 5.53

If we want to glue together two opposite edges of the square $[0, 1] \times [0, 1]$, then we would define an equivalence relation

$$(x, y) \sim (x', y') \iff x = x' \text{ and } y - y' \in \mathbf{Z}.$$

In other words, only glue (x, y) to (x', y') if the x-coordinates agree and the y-coordinates either agree, or one is 0 and the other is 1.

The result, in this case, should be homeomorphic to a cylinder:

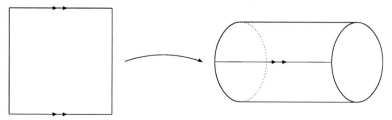

Once we have an equivalence relation on a set, we can form a quotient set, $X/\sim$, with a natural function $\pi : X \to X/\sim$ taking an element $x \in X$ to its equivalence class. As above, the topology we put on the **quotient space** $X/\sim$ is the one with as many open sets as possible, subject to the condition that π be continuous. In other words, a subset $U \subset X/\sim$ is open if $\pi^{-1}(U)$ is open.

Example 5.54

With this topology, the quotient $[0, 1] \times [0, 1]/\sim$ of Example 5.53 is homeomorphic to the product $[0, 1] \times ([0, 1]/\partial[0, 1])$ which, in turn, is homeomorphic to the cylinder $[0, 1] \times S^1$, as expected.

Example 5.55

If we reverse the orientation by putting on the relation

$$(x, y) \sim (x', y') \iff \text{ either } (x, y) = (x', y') \text{ or } x = 1 - x' \text{ and } y - y' = \pm 1,$$

then we obtain the Möbius band.

Example 5.56

We can glue together both pairs of opposite edges of the square, by the equivalence relation

$$(x, y) \sim (x', y') \iff x - x' \in \mathbf{Z} \text{ and } y - y' \in \mathbf{Z}.$$

This gives a space homeomorphic to the product $S^1 \times S^1$ and, hence, homeomorphic to the torus T^2.

Example 5.57

Alternatively, we can glue together both pairs of opposite edges of the square, but with the orientation reversed on one pair, by defining

$$(x, y) \sim (x', y') \iff \text{ either } x - x' \in \mathbf{Z} \text{ and } y = y' \text{ or } x = 1 - x' \text{ and } y - y' = \pm 1.$$

This gives a space we have not met before, the **Klein bottle**. This cannot be embedded in three-dimensional space, so it is hard to visualize. The following picture shows a three-dimensional image of the Klein bottle, achieved by allowing some intersections which do not occur in the Klein bottle itself.

Like the torus, we can think of this as a cylinder with the boundary circles glued together, so that the arrows in the diagram match up.

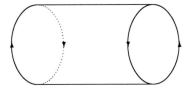

Note that the arrows on the right-hand circle are in the opposite direction to the way they would be if we were making a torus.

Like the Möbius band, the Klein bottle has only one side. In fact, if we cut the Klein bottle in two pieces, along the horizontal plane through the arrowheads in the above diagram, then we get two Möbius bands.

Example 5.58

If we take $\mathbf{R}^n - \{0\}$ and put on the equivalence relation

$$\mathbf{x} \sim \mathbf{y} \iff \lambda\mathbf{x} = \mu\mathbf{y} \text{ for some } \lambda, \mu \in \mathbf{R} - \{0\},$$

then we get a space which is homeomorphic with the real projective space $\mathbf{RP}^{n-1}$. In fact, if you look back at how we defined the topology on $\mathbf{RP}^{n-1}$, you can see that it is exactly the same as the construction of the topology on this quotient space, phrased in different language.

Example 5.59

We can also describe $\mathbf{RP}^{n-1}$ as a quotient of S^{n-1}, under the equivalence relation

$$\mathbf{x} \sim \mathbf{y} \iff \mathbf{x} = \pm\mathbf{y}.$$

Example 5.60

Since S^1 is homeomorphic to $\mathbf{RP}^1$, the previous example shows that S^1 is, in fact, homeomorphic to a quotient of itself: $S^1 \cong S^1/\sim$ where the equivalence relation is defined by

$$\mathbf{x} \sim \mathbf{y} \iff \mathbf{x} = \pm\mathbf{y}.$$

For example, a homeomorphism $S^1/\sim \to S^1$ is given by $z \mapsto z^2$.

Note that the concept of a quotient $X/\sim$ of a space by an equivalence relation completely subsumes the earlier notion of the quotient X/A of a space by a subspace:

Proposition 5.61

If A is a subspace of the topological space X, then we can describe the quotient X/A as $X/\sim$ where the equivalence relation is defined by

$$x \sim y \iff x, y \in A, \quad \text{or} \quad x = y.$$

Hence, if we write about a "quotient space $X/\sim$", then anything we say also applies to quotients X/A. In particular, the following description of continuous maps defined on $X/\sim$ in terms of continuous maps defined on X also serves to describe maps on X/A.

Theorem 5.62

Suppose that X is a topological space with an equivalence relation $\sim$ defined on it, and Q is the quotient space $Q = X/\sim$. If S is any other topological space and $f : Q \to S$ is a continuous function, then we can compose f with the projection $\pi : X \to Q$ to get a function $X \to S$ which will also be continuous. In other words, given a function f as in the diagram, there is a function $f \circ \pi : X \to S$, as shown by a dotted line, such that the diagram commutes:

i.e., whichever route you take from X to S, you get the same function $f \circ \pi$.

Conversely, if $g : X \to S$ is any continuous function such that $g(x) = g(y)$ whenever $x \sim y$, then there is a continuous function $f : Q \to S$ such that $f \circ \pi = g$, i.e., the following diagram commutes

Proof

The first part of the theorem is a simple example of Proposition 3.10; $f \circ \pi$ is continuous because both f and π are continuous.

Conversely, if $g : X \to S$ is continuous and $g(x) = g(y)$ whenever $x \sim y$, then we can derive a map $f : Q \to S$ by $f(E) = g(x)$ if E is an equivalence class and x a member of E. The key point is that if y is another member of E, then $x \sim y$, so $g(x) = g(y)$ and $f(E) = g(y)$. Thus it does not matter which member of E we choose, as g gives the same value on each of them. (We say that f is **well defined**, meaning that the original function g respects the equivalence relation, and so defines a function f on the set of equivalence classes.) And we certainly have $f \circ \pi = g$.

To show that this map f is continuous, let R be an open subset of S. Then

$$g^{-1}(R) = (f \circ \pi)^{-1}(R) = \pi^{-1}(f^{-1}(R)).$$

The function g is continuous, so $g^{-1}(R)$ is open, i.e., $\pi^{-1}(f^{-1}(R))$ is open. And the topology on Q was defined so that the preimage π^{-1} of a set is open precisely when the set is open. So $\pi^{-1}(f^{-1}(R))$ being open means that $f^{-1}(R)$ is open. Hence f is continuous.   □

Corollary 5.63

The constructions in Theorem 5.62 give a one-to-one correspondence between continuous maps $f : (X/\sim) \to S$ and continuous maps $g : X \to S$ such that $g(x) = g(y)$ whenever $x \sim y$.

Example 5.64

We can describe the set of continuous functions from the real projective space $\mathbf{RP}^3$ to the torus T^2 as follows. Using the fact that $\mathbf{RP}^3$ is the quotient of S^3 obtained by identifying opposite points, and T^2 is the product $S^1 \times S^1$, we see that a map $\mathbf{RP}^3 \to T^2$ corresponds to a pair of maps $f, g : S^3 \to S^1$ which satisfy $f(\mathbf{x}) = f(-\mathbf{x})$ and $g(\mathbf{x}) = g(-\mathbf{x})$. If we can describe the set of continuous functions between two innocuous spaces such as S^3 and S^1, then we can derive a description of the continuous functions $\mathbf{RP}^3 \to T^2$.

We can also deduce something about topological properties of the quotient from properties of the original space.

Proposition 5.65

If $Q = X/\sim$ and X is compact, then Q is compact.

Proof

The quotient construction always gives rise to a surjection $X \to X/\sim$ which, by definition of the quotient topology, is continuous. The result then follows from Proposition 4.27.   □

Example 5.66

Since real projective space $\mathbf{RP}^n$ can be thought of as a quotient of S^n (Example 5.59), this shows that $\mathbf{RP}^n$ is compact.

This helps us to prove a rather surprising homeomorphism.

Proposition 5.67

The space $SO(3)$ of special orthogonal 3×3 matrices is homeomorphic with $\mathbf{RP}^3$.

Proof

The map $S^3 \to SO(3)$ of Example 3.41 takes antipodal points to the same matrix, i.e., $(w, x, y, z) \in S^3$ and $(-w, -x, -y, -z)$ have the same image. Hence this descends to a map $\mathbf{R}P^3 \to SO(3)$. This is then an injection and, in fact, a bijection, since the map $S^3 \to SO(3)$ is surjective. Now $\mathbf{R}P^3$ is compact and $SO(3)$ is Hausdorff, being a subspace of the Hausdorff space $\mathbf{R}^9$. Hence the map $\mathbf{R}P^3 \to SO(3)$ is a homeomorphism by Theorem 5.24.   □

Unlike compactness, the Hausdorff property does not pass to quotients: $X/\sim$ need not be Hausdorff even if X is.

Example 5.68

The real line with a double point that we met in Example 4.37 can be described as a quotient of $\mathbf{R} \times S^0$. We put on the equivalence relation

$$(x, t) \sim (x', t') \iff \text{either } x = x' \neq 0 \text{ or } (x, t) = (x', t').$$

Although $\mathbf{R} \times S^0$ is Hausdorff, being a product of two Hausdorff spaces, this quotient is not, as demonstrated in Example 4.37.

This example also shows that $X/\sim$ may be Hausdorff even if X is not, since $\mathbf{R}$ (which is Hausdorff) is the quotient of this real line with a double point under the equivalence relation that identifies the two 0's together.

Quotient spaces do, however, respect connectivity.

Proposition 5.69

If X is connected and $X/\sim$ is a quotient space of X, then $X/\sim$ is also connected.

Proof

By definition of the term "quotient space", there is a continuous surjection $X \to X/\sim$. Hence, by Proposition 4.11, if X is connected, then so is $X/\sim$.   □

Example 5.70

Since S^n is connected for $n > 0$, this shows that $\mathbf{R}P^n$ is connected for $n > 0$.

The converse is not true: $X/\sim$ may be connected even though X is disconnected, as the following example shows.

Example 5.71

Let $X = [0, 1/2] \amalg [1/2, 1]$. Because we have taken a *disjoint* union, this contains one point for every $t \in [0, 1]$ except for $t = 1/2$, for which there are two points, which we label L and R. We then define an equivalence relation on X by

$$x \sim y \iff (x \text{ is either } L \text{ or } R) \text{ and } (y \text{ is either } L \text{ or } R) \text{ or } x = y.$$

In other words, the only points we identify together are the two points L and R which correspond to $t = 1/2$. Then $X/\sim$ is homeomorphic to $[0, 1]$. To see this, we first define $\tilde{f} : X \to [0, 1]$ by $\tilde{f}(t) = t$. This is continuous because $\tilde{f}^{-1}(a, b) = ((a, b) \cap [0, 1/2]) \amalg ((a, b) \cap [1/2, 1])$, a disjoint union of an open set of $[0, 1/2]$ and an open set in $[1/2, 1]$. Moreover, $\tilde{f}(L) = \tilde{f}(R) = 1/2$, so $\tilde{f}$ corresponds to a continuous function $f : X/\sim \to [0, 1]$. Clearly f is bijective; its inverse $g : [0, 1] \to X/\sim$ is given by $g(t) = t$ if $t \neq 1/2$, $g(1/2) = L = R$.

This map g is continuous, because if $U \subset X/\sim$ is open, then $\pi^{-1}(U)$ is an open subset of X, i.e., a disjoint union of an open subset of $[0, 1/2]$ and an open subset of $[1/2, 1]$. We may assume this disjoint union is of the form

$$((a, b) \cap [0, \frac{1}{2}]) \amalg ((c, d) \cap [\frac{1}{2}, 1])$$

since the open intervals form a basis for the topology on $\mathbf{R}$. The preimage $g^{-1}(U)$ is then $((a, b) \cup (c, d)) \cap [0, 1]$, which is an open set in $[0, 1]$. Hence g is continuous, so f and g are homeomorphisms.

It seems perverse to deconstruct the interval in this way, but it leads, via Theorem 5.62, to the following

Lemma 5.72

If S is any topological space and $f : [0, 1/2] \to S$, $g : [1/2, 1] \to S$ are two continuous maps such that $f(1/2) = g(1/2)$, then they can be combined to give a continuous map $h : [0, 1] \to S$ such that

$$h(t) = \begin{cases} f(t) & \text{if } t \leq \frac{1}{2}, \\ g(t) & \text{if } t \geq \frac{1}{2}. \end{cases}$$

In other words, if we have two maps each defined on half of the interval $[0, 1]$ such that they agree where they are both defined, then we get a continuous map on the whole of the interval. This can be generalized in the following way.

Lemma 5.73 (Gluing Lemma)

Let T be any topological space, with closed subsets $U_1, \ldots, U_n$ such that every point in T belongs to at least one of the subsets $U_1, \ldots, U_n$. If S is any topological space and there are maps $f_i : U_i \to S$ for each $i = 1, \ldots, n$ which are continuous with respect to the subspace topology on U_i, and such that $f_i(x) = f_j(x)$ for all $x \in U_i \cap U_j$, then they can be glued together to give a continuous map $f : T \to S$ such that

$$f(x) = f_i(x) \quad \text{if } x \in U_i.$$

Proof

We will show that the space T is homeomorphic with the quotient $Q = (U_1 \amalg U_2 \amalg \cdots \amalg U_n)/\sim$ where the equivalence relation is defined by $x \sim y$ if $x \in U_i$ and $y \in U_j$ and $h_i(x) = h_j(y)$, where $h_i : U_i \to T$, $h_j : U_j \to T$ are the natural inclusion maps. To see this, let $D = U_1 \amalg U_2 \amalg \cdots \amalg U_n$ and define

$$g : D \to T$$

by $g(x) = h_i(x)$ if $x \in U_i$. This respects the equivalence relation so, by Theorem 5.62, gives a continuous function $\bar{g} : Q \to T$. This function $\bar{g}$ is bijective since the sets $U_1, \ldots, U_n$ cover T, so $\bar{g}$ has an inverse function $\bar{g}^{-1} : T \to Q$. To show that this inverse function is continuous, we will use Lemma 5.27, so let V be a closed subset of Q. Because $\bar{g}$ is bijective, the preimage of V under $\bar{g}^{-1}$ is the image, under g, of V. By construction this is the union of $U_i \cap V$ for all i. Since U_i and V are closed, $U_i \cap V$ is also closed, and the finite union $\cup_{i=1}^{n}(U_i \cap V)$ is closed. Hence $g(V)$ is closed, i.e., the preimage of the closed set V, under the map $\bar{g}^{-1}$, is closed. Hence $\bar{g}^{-1}$ is continuous, and $\bar{g}$ is a homeomorphism as claimed. □

So we can define a map piece by piece, and, so as long as the pieces are defined on closed subsets and agree on any overlap that there may be, they can be glued together to form a continuous map on the whole space. This is tremendously useful in constructing maps.

EXERCISES

5.1. Prove that $[1, 2)$ is homeomorphic to $(-1, 0]$.

5.2. Let $S = \{a, b, c, d\}$ with the discrete topology and let $T = \{a, b, c, d\}$ with the indiscrete topology. Define $f : S \to T$ to be the identity map. Is f a homeomorphism ?

5.3. Verify that the map $f : (-1,1) \to \mathbf{R}$ of Example 5.2, given by $f(x) = \tan(\pi x/2)$, is continuous.

5.4. Verify that the map $g : S^1 \to \mathbf{RP}^1$ of Example 5.6 is continuous.

5.5. Prove that the stereographic projection $S^1 - \{(0,1)\} \to \mathbf{R}$ of Example 5.7 is given by the formula $(x,y) \mapsto 2x/(1-y)$, and derive the formula for the projection $S^2 - \{(0,0,1)\} \to \mathbf{R}^2$.

5.6. Show that the cube $[0,1]^3 = \{(x,y,z) \in \mathbf{R}^3 : 0 \le x,y,z \le 1\}$ is homeomorphic with the solid 3-sphere

$$\{(x,y,z) \in \mathbf{R}^3 : x^2 + y^2 + z^2 \le 1\}.$$

5.7. Show that the annulus $A = \{(x,y) \in \mathbf{R}^2 : 1 \le x^2 + y^2 \le 4\}$ is homeomorphic to the cylinder $C = \{(x,y,z) \in \mathbf{R}^3 : x^2 + y^2 = 1, 0 \le z \le 1\}$.

5.8. Prove that $[0,1] \amalg [1,2]$ is homeomorphic to $S^0 \times [0,1]$.

5.9. Prove that $S \times T$ is Hausdorff if S and T are both Hausdorff. Prove that $S \times T$ is compact only if S and T are both compact. Prove that $S \times T$ is connected only if S and T are both connected.

5.10. Prove that the quotient $S^1 \times [0,1]/\sim$ is homeomorphic to the sphere S^2, where $((x,y),t) \sim ((x',y'),t')$ if either $t = t' = 1$, $t = t' = 0$ or $((x,y),t) = ((x',y'),t')$.

5.11. The following two ways of gluing together the edges of an octagon give rise to quotient spaces homeomorphic to two different spaces from Chapter 3. What are they?

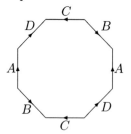 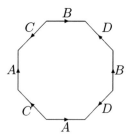

Interlude

As we have seen, topological properties can tell us a fair amount of information about continuous maps. Nevertheless, they are not very discerning. For each property, either a space has the property or it does not; it's black or white. This simply groups spaces into two classes: Those with the property and those without it.

There are, however, other topological invariants, which are much more discerning. For example, the Euler number, which we will meet in Chapter 7, assigns an integer to each space. If the Euler number is 0, it tells us one thing, while if it is 1, it tells us something different, and if it is -2, it tells us something else again, and so on. This carries much more information about the space and, as a result, can tell us much more about continuous maps than a simple property can.

In Chapters 7–10 we will meet a number of different topological invariants. These have proved to be tremendously important over the last 100 years, and finding new invariants has become one of the major themes of research in topology.

All of the invariants covered in these chapters have one limitation. They are all blind to one aspect of topology, which is that they do not distinguish between "homotopic" functions, i.e. functions which can be continuously deformed into each other. We will examine this notion of homotopy in Chapter 6, where we will see that this is less limiting than it first appears.

After introducing and studying these invariants, we will finish, in Chapter 11, by considering more sophisticated deconstructive techniques which are particularly appropriate for understanding these invariants.

6
Homotopy

We said at the beginning of this book that topology is about the study of continuous functions and so the ultimate goal of topology should be to describe all the continuous maps between any given pair of topological spaces. Of course, with almost any pair of spaces, there are lots of continuous functions between them – far more than we can ever hope to list or understand. For example, it is not remotely feasible to list even the continuous functions from the interval $[0, 1]$ to itself.

However, if we allow some leeway, then this difficulty can be avoided. The idea is that we should consider two functions to be equivalent, or "homotopic", if one can be deformed into the other.

6.1 Homotopy

For example, let $f : [0, 2] \to \mathbf{R}$ be the function $f(x) = 1 + x^2(x - 2)^2$, depicted below.

This is almost a constant function to 1, but with a small deviation around $x = 1$. If we take the function $f_1(x) = 1 + \frac{1}{2}x^2(x - 2)^2$, then this has a similar

M.D. Crossley, *Essential Topology*, Springer Undergraduate
Mathematics Series, DOI 10.1007/978-1-84628-194-5_6,
© Springer-Verlag London Limited 2010

shape, but with a smaller deviation. Similarly, $f_2(x) = 1 + \frac{1}{3}x^2(x-2)^2$ has the same shape but with an even smaller deviation.

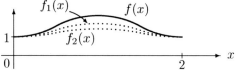

Carrying on, for each $n \geq 1$, we can define $f_n(x) = 1 + \frac{1}{(n+1)}x^2(x-2)^2$, and thus obtain a family of functions interpolating between f and the constant function.

However, we need these interpolating functions to provide a *continuous* deformation of the one function into the other. To achieve this, we should not parametrize the interpolating functions f_1, f_2, etc. by integers, but, instead, we should index them by real numbers in some fixed range, say between 0 and 1. So we would then want a family of functions $\{f_t\}_{t\in[0,1]}$, such that $f_0 = f$, and f_1 is the constant function to 1.

In the above example, we can set $f_t(x) = 1 + (1-t)x^2(x-2)^2$ for each $t \in [0,1]$. Then $f_0(x) = 1 + x^2(x-2)^2 = f(x)$ and $f_1(x) = 1$ is the constant function.

Such a deformation then assigns a function to each point in $[0,1]$, so the deformation is a function from $[0,1]$ to the set of continuous maps $[0,2] \to \mathbf{R}$, which takes $t \in [0,1]$ to the function f_t. For the deformation to be continuous, we should obviously ask that this function be continuous. However, this would require us to put a topology on the set of continuous maps $[0,2] \to \mathbf{R}$ and, more generally, on the set of maps $S \to T$ for any topological spaces S and T. This can be done, and we will see how in Chapter 11, but for now we will use a simpler route to specify that the deformation be continuous.

Note that the family $\{f_t\}_{t\in[0,1]}$ assigns, to each point $t \in [0,1]$, a function $f_t : [0,2] \to \mathbf{R}$. This, in turn, assigns to each point $x \in [0,2]$, a value $f_t(x) \in \mathbf{R}$. Thus we can think of this family as assigning to each pair $(x,t) \in [0,2] \times [0,1]$ the value $f_t(x) \in \mathbf{R}$. In other words, we have a function $[0,2] \times [0,1] \to \mathbf{R}$. Since we have a topology on $[0,2]$, and we know a topology on $[0,1]$, we can use the product topology to topologize $[0,2] \times [0,1]$, and therefore our interpolating family corresponds to a function between two topological spaces. Thus, we can define the family to be continuous if the corresponding function is continuous. Hence we arrive (finally!) at the following definition.

Definition: Two maps $f, g : S \to T$ are **homotopic** if there is a continuous function

$$F : S \times [0,1] \longrightarrow T$$

such that $F(s,0) = f(s)$ for all $s \in S$ and $F(s,1) = g(s)$ for all $s \in S$. In this case, F is a **homotopy** between f and g, and we write $f \simeq g$.

Example 6.1

In the preceding example, where $f : [0, 2] \to \mathbf{R}$ is given by $f(x) = 1 + x^2(x-2)^2$, the function $F : [0, 2] \times [0, 1] \to \mathbf{R}$ given by $F(x, t) = 1 + (1-t)x^2(x-2)^2$ is continuous, being a polynomial, and satisfies $F(x, 0) = 1 + x^2(x-1)^2 = f(x)$ and $F(x, 1) = 1$. Thus F is a homotopy from f to the constant function to 1.

Example 6.2

Let $f : S^1 \to \mathbf{R}^2$ be the natural inclusion map $f(x, y) = (x, y)$, and let $g : S^1 \to \mathbf{R}^2$ be the constant map $g(x, y) = (0, 0)$ for all $(x, y) \in S^1$. These two maps are homotopic, for the function $F : S^1 \times [0, 1] \to \mathbf{R}^2$ defined by

$$F((x, y), t) = (1 - t)f(x, y)$$

is continuous, and has the property that $F((x, y), 0) = (1 - 0)f(x, y) = f(x, y)$ and $F((x, y), 1) = (1 - 1)f(x, y) = (0, 0) = g(x, y)$.

Since the interval $[0, 1]$ features in the definition, it is very prevalent in discussions of homotopy. It is useful, therefore, to refer to it simply as I. This also serves to emphasize its rôle as the homotopy parameter, as in the following example.

Example 6.3

Let $f : [0, 1] \to [0, 1]$ be the identity map and let $g : [0, 1] \to [0, 1]$ be the constant map $g(x) = 0$ for all x. Then there is a homotopy $F : [0, 1] \times I \to [0, 1]$ between these maps given by

$$F(x, t) = (1 - t)x.$$

Example 6.4

Let $f, g : \mathbf{R} \to \mathbf{R}$ be any two continuous functions. Define

$$F : \mathbf{R} \times I \to \mathbf{R}$$

by $F(x, t) = (1 - t)f(x) + tg(x)$. Then F is continuous, being a composite of continuous functions, $F(x, 0) = (1-0)f(x)+0 = f(x)$ and $F(x, 1) = 0+1g(x) = g(x)$, so F is a homotopy between f and g. In other words, any two continuous functions on $\mathbf{R}$ are homotopic.

This idea can be used with any "convex" range space. A subspace T of $\mathbf{R}^n$ is said to be **convex** if, given any two points x, y in T, the straight line from x to y is contained in T. In other words, for any number $t \in I$, the point $tx + (1 - t)y$ is in T.

Proposition 6.5

If T is convex, and S is any topological space, then any two maps $f, g : S \to T$ are homotopic.

Proof

Define the homotopy $F : S \times I \to T$ by

$$F(x,t) = tf(x) + (1 - t)g(x). \qquad \Box$$

On the other hand, we cannot use this argument for maps to S^1, for example, since if $f(x)$ and $g(x)$ are two distinct points in S^1, then $tf(x) + (1 - t)g(x)$ will not usually be a point in S^1, as depicted below:

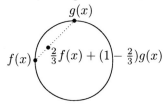

Now, we want to consider homotopic functions to be "the same". In other words, we want to form equivalence classes of homotopic functions, for which we need the following three lemmas.

Lemma 6.6

Let $f : S \to T$ be any continuous map. Then $f \simeq f$.

Proof

We can define a homotopy

$$F : S \times I \to T$$

by $F(x,t) = f(x)$ for all t. Then $F(x,0) = f(x)$ and $F(x,1) = f(x)$. $\qquad \Box$

Lemma 6.7

Let $f, g : S \to T$ be two continuous maps. If F is a homotopy between f and g, then there is also a homotopy between g and f.

Proof

If $F(x,0) = f(x)$ and $F(x,1) = g(x)$, then define

$$G : S \times I \to T$$

by $G(x,t) = F(x,1-t)$. Then $G(x,0) = g(x)$ and $G(x,1) = f(x)$. □

Lemma 6.8

Let $f, g, h : S \to T$ be three continuous maps. If f and g are homotopic and g and h are homotopic, then f and h are homotopic.

Proof

Let $F : S \times I \to T$ be a continuous map such that $F(x,0) = f(x)$ and $F(x,1) = g(x)$, and let $G : S \times I \to T$ be a continuous map such that $G(x,0) = g(x)$ and $G(x,1) = h(x)$. Define a function $H : S \times I \to T$ by

$$H(x,t) = \begin{cases} F(x,2t) & \text{if} \quad 0 \le t \le \frac{1}{2}, \\ G(x,2t-1) & \text{if} \quad \frac{1}{2} \le t \le 1. \end{cases}$$

This is continuous by the gluing lemma, and has the property that $H(x,0) = F(x,0) = f(x)$, while $H(x,1) = G(x,1) = h(x)$. □

These results say that we can form a set of equivalence classes of homotopic functions between two given topological spaces. We write $[S,T]$ for the set of **homotopy classes** of maps $S \to T$. This is much more manageable than the complete set of continuous maps from S to T. For example, if $S = T = \mathbf{R}$, then, by Example 6.4, all functions $S \to T$ are homotopic, so $[\mathbf{R}, \mathbf{R}]$ consists of a single element. This is an extreme case. A more interesting example is that, as we shall see later, $[S^1, S^1]$ contains one element for each integer. This gives some reason to believe that homotopy classes of maps still contain some information about the topology of the spaces involved.

Of course, for homotopy classes to be a useful tool in studying continuous functions, they must respect the most basic operation on functions, namely composition. Fortunately, they do:

Proposition 6.9

If $f \simeq g : S \to T$ and $h \simeq j : T \to U$, then $(h \circ f) \simeq (j \circ g) : S \to U$.

Proof

Let $F : S \times [0,1] \to T$ be a homotopy from f to g, and let $H : T \times [0,1] \to U$ be a homotopy from h to j. Define a homotopy $G : S \times I \to U$ by

$$G(s,t) = H(F(s,t),t).$$

It is straightforward to check that $G(s,0) = h(f(s))$ and $G(s,1) = j(g(s))$, and the map G is continuous since it is a composite of continuous maps.   □

6.2 Homotopy Equivalence

If we are going to consider two functions to be equivalent when they are homotopic, then we should modify the definition of homeomorphism, replacing the $=$ signs by homotopies. This leads to the following notion of "homotopy equivalence":

Definition: Two topological spaces S, T are said to be **homotopy equivalent** if there are continuous maps $f : S \to T$ and $g : T \to S$ such that $g \circ f$ is homotopic to the identity on S and $f \circ g$ is homotopic to the identity on T. The maps f and g are then **homotopy equivalences**. If S and T are homotopy equivalent, then we write $S \simeq T$.

Lemma 6.10

If $S \simeq T$ and Q is any topological space, then $[S,Q] = [T,Q]$ and $[Q,S] = [Q,T]$.

Proof

If $S \simeq T$, then there are maps $f : S \to T$, $g : T \to S$ whose composites are homotopic to the respective identity maps. Now if $h : S \to Q$, then we can compose with g to obtain a map $(h \circ g) : T \to Q$, and if $j : T \to Q$, then we can compose with f to obtain a map $(j \circ f) : S \to Q$. And, up to homotopy, these two operations are mutually inverse: $(h \circ g) \circ f = h \circ (g \circ f) \simeq h \circ 1_S = h$, while $(j \circ f) \circ g = j \circ (f \circ g) \simeq j \circ 1_T = j$.

In a similar way, by composing with f or with g we can get correspondences between $[Q,S]$ and $[Q,T]$.   □

At the time of writing, it is becoming increasingly common to say that two spaces are **homotopic** rather than "homotopy equivalent".

Lemma 6.11

If S and T are homeomorphic, then they are also homotopy equivalent.

Proof

If we have homeomorphisms $f : S \to T$ and $g : T \to S$, then $f \circ g$ and $g \circ f$ are the respective identity maps, so these composites are homotopic to the respective identity maps by Lemma 6.6. □

Of course, there are many pairs of spaces which are homotopy equivalent but not homeomorphic.

Example 6.12

If S is a space containing a single point, then S and $\mathbf{R}$ are homotopy equivalent. To see this, define $f : \mathbf{R} \to S$ to be the constant function (there is no choice as to how to define f), and let $g : S \to \mathbf{R}$ be the function which takes the single point in S to 0 in $\mathbf{R}$. The composite $f \circ g : S \to S$ is the identity map, while the composite $g \circ f : \mathbf{R} \to \mathbf{R}$ is the constant function to 0. Since all functions $\mathbf{R} \to \mathbf{R}$ are homotopic, by Example 6.4, so $g \circ f$ is homotopic to the identity.

By Lemma 6.10, this tells us that $[\mathbf{R}, \mathbf{R}] = [\{0\}, \{0\}]$. Since there is only one continuous function $\{0\} \to \{0\}$, there can be only one homotopy class of maps $\{0\} \to \{0\}$. Thus $[\{0\}, \{0\}]$ contains only one element and, consequently, so does $[\mathbf{R}, \mathbf{R}]$, confirming Example 6.4.

A space which, like $\mathbf{R}$, is homotopy equivalent to a one-point space is said to be **contractible**.

Example 6.13

The interval $[0, 1]$ is homotopy equivalent to a one-point space $\{0\}$: Define $f : [0, 1] \to \{0\}$ by $f(x) = 0$, and define $g : \{0\} \to [0, 1]$ by $g(0) = 0$. Then $(f \circ g) : \{0\} \to \{0\}$ is the identity map, and so this is certainly homotopic to the identity map.

Conversely, $(g \circ f)(x) = 0$ for all x. This is homotopic to the identity map by Example 6.3.

Example 6.14

In the same way we can show that the open interval $(0, 1)$ is homotopy equivalent to $\{0\}$. Of course, we need to define $g : \{0\} \to (0, 1)$ differently, for example by $g(0) = 1/2$. With this choice of g, a homotopy from $(g \circ f)$ to the identity map of $(0, 1)$ is given by $H : (0, 1) \times I \to (0, 1)$ defined by

$$H(x, t) = \frac{1 - t}{2} + tx.$$

The image $H(x, t)$ is certainly contained in $(0, 1)$ if $(x, t) \in (0, 1) \times I$, and H is continuous, being a composite of multiplications and additions.

Example 6.15

Consequently, any open interval (a, b) is homotopy equivalent to $\{0\}$, since (a, b) is homeomorphic with $(0, 1)$. This applies even to infinite intervals (a, ∞) and $(-\infty, b)$.

Proposition 6.16

If S is contractible and T is any topological space, then any two continuous functions $f, g : T \to S$ are homotopic. In particular, any continuous function to a contractible space is homotopic to a constant map.

Proof

Let $f, g : T \to S$ be two continuous maps. If S is contractible, then there are continuous maps $h : S \to \{0\}$ and $j : \{0\} \to S$ such that $h \circ j \simeq 1$ and $j \circ h \simeq 1$. In particular,

$$f = (1 \circ f) \simeq (j \circ h \circ f) \quad \text{and} \quad g = (1 \circ g) \simeq (j \circ h \circ g).$$

Since $h \circ f : T \to \{0\}$, so $j \circ h \circ f : T \to S$ must be the constant map $t \mapsto j(0)$ for all $t \in T$. Similarly, $j \circ h \circ g$ is this same constant map, and so $f \simeq g$. $\qquad\square$

Of course, there are many pairs of spaces which are homotopy equivalent without being contractible.

Example 6.17

Let A be the annulus

$$A = \{(x, y) \in \mathbf{R}^2 : 1 \le \sqrt{x^2 + y^2} \le 2\}.$$

Then $A \simeq S^1$ as follows. Define $f : S^1 \to A$ to be the natural inclusion $f(x, y) = (x, y)$, and $g : A \to S^1$ to be the radial projection inwards

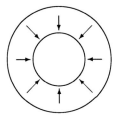

which can be described algebraically as

$$g(x, y) = \frac{1}{\sqrt{x^2 + y^2}}(x, y).$$

Now, $g \circ f$ is the identity on S^1 because, if $(x, y) \in S^1$, then $g(x, y) = (x, y)$. This is certainly homotopic to the identity, by Lemma 6.6.

And $(f \circ g)(x, y) = \left(1/\sqrt{(x^2 + y^2)}\right)(x, y)$. This is homotopic to the identity on A by the homotopy $F : A \times I \to A$ defined by

$$F((x, y), t) = \frac{t\sqrt{x^2 + y^2} + (1 - t)}{\sqrt{x^2 + y^2}}(x, y).$$

Check: This is continuous, being a composite of continuous maps; $F((x, y), 0) = (f \circ g)(x, y)$ and $F((x, y), 1) = (x, y)$.

Hence f and g form a homotopy equivalence between A and S^1. We will see in the next section that these spaces are not contractible.

Example 6.18

Similarly, the space $\mathbf{C}^{\times} = \mathbf{R}^2 - \{(0, 0)\}$ is homotopy equivalent to S^1.

Proving that two spaces are not homotopy equivalent is hard, just as it was hard to prove directly that two spaces are not homeomorphic. One case where we can do this is when we are dealing with finite discrete spaces, such as S^0.

Proposition 6.19

The 2-point space S^0 is not contractible.

Proof

Suppose that S^0 is contractible, with homotopy equivalences $f : S^0 \to \{0\}$ and $g : \{0\} \to S^0$. Then $f \circ g : \{0\} \to \{0\}$ has to be the identity, and $g \circ f : S^0 \to S^0$

is homotopic to the identity. So there is a homotopy

$$F : S^0 \times I \to S^0$$

with $F(x, 0) = x$ and $F(x, 1) = g(f(x)) = g(0)$.

Now define a map $h : I \to S^0$ by

$$h(t) = F(-g(0), t).$$

This will be a continuous map, with $h(0) = F(-g(0), 0) = -g(0)$ and $h(1) = F(-g(0), 1) = g(0)$. Since S^0 only has two points, h must be surjective. But Lemma 4.3 and Example 4.5 show that this cannot happen. Hence there cannot have been a homotopy equivalence between S^0 and $\{0\}$. $\square$

This idea can be developed to show that a discrete space consisting of m points is homotopy equivalent to a discrete space consisting of n points *only* if $m = n$. It can also be developed to show that a connected space cannot be homotopy equivalent to a disconnected space.

Proposition 6.20

If X is connected and Y is disconnected, then X and Y are not homotopy equivalent.

Proof

Suppose that X and Y are homotopy equivalent, with maps $f : X \to Y$ and $g : Y \to X$ whose composites are homotopic to the identity. In particular, there is a homotopy $F : Y \times I \to Y$ such that $F(y, 0) = f(g(y))$ and $F(y, 1) = y$ for all $y \in Y$.

If Y is disconnected, then it can be expressed as a disjoint union $Y = U \amalg V$ where both U and V are open and non-empty, and so there is a continuous surjection $p : Y \to S^0$ with $p(y) = 1$ if $y \in U$, $p(y) = -1$ if $y \in V$.

Since X is connected, the map f has image contained in one of these components, say Im $f \subset U$. Since V is not empty, there is at least one point $v \in V$, and we can define a map $h : [0, 1] \to S^0$ by

$$h(t) = p(F(v, t)).$$

Since $F(v, 0) = f(g(v)) \in U$ (because Im $f \subset U$) and $F(v, 1) = v \in V$, so $h(0) = 1$ and $h(1) = -1$. Thus h is a surjection $[0, 1] \to S^0$, and h is continuous as it is a composite of continuous maps. As in the preceding proposition, this is not possible, so X and Y cannot be homotopy equivalent. $\square$

Thus we can use connectivity to distinguish homotopy inequivalent spaces.

Example 6.21

The circle S^1 is not homotopy equivalent to the 0-sphere S^0.

Unfortunately, compactness cannot be used in this way since Examples 6.13 and 6.14 exhibit two spaces which are both contractible and, hence, homotopy equivalent, but one is compact and the other is not. Similarly, there are Hausdorff spaces which are homotopy equivalent to non-Hausdorff spaces.

Example 6.22

Let $S = \{1, 2\}$ with the indiscrete topology, as in Example 4.36. Let T be a one-point space, say $T = \{0\}$. We can define $f : T \to S$ by $f(0) = 1$ and this is continuous since every function to an indiscrete space is continuous (Proposition 3.9). We can define $g : S \to T$ by $g(s) = 0$, this being continuous as T is also indiscrete. Then $g \circ f : T \to T$ is the identity map, and $f \circ g : S \to S$ is the constant map to 1. This is not the identity, but is homotopic to it, as we will now show. Such a homotopy will be a function $F : S \times I \to S$. Since S is indiscrete, any such function is continuous, i.e., we can define F any way we choose and it will be continuous. In particular, we can define F by

$$F(s, t) = \begin{cases} s & \text{if} \quad t \leq \frac{1}{2}, \\ 1 & \text{if} \quad t > \frac{1}{2}. \end{cases}$$

Hence $F(s, 0) = s$ is the identity on S, and $F(s, 1) = 1 = (f \circ g)(s)$. Thus we have a homotopy from $f \circ g$ to the identity, so S is homotopy equivalent to T. Example 4.36 showed that S is not Hausdorff, whereas T is Hausdorff, being a subspace of $\mathbf{R}$.

So the properties developed in Chapter 4 are of limited use in a homotopy context. In particular, there are many interesting spaces which share all of these properties, while being quite distinct. For example the circle, S^1, is connected, compact, and Hausdorff, i.e., it looks just like a point as far as Chapter 4 is concerned. Instinctively, we can see that S^1 is not homotopy equivalent to a point, but our instinct can sometimes be wrong (our instinct would tell us that $\{1, 2\}$ could not be homotopy equivalent to a one-point space, in contrast to Example 6.22) so we need a rigorous proof before we can be entirely confident. The next section contains such a proof.

6.3 The Circle

We wish to prove that the circle is not contractible. However, for only a little extra effort, we can perform a more impressive calculation which will enable us to list all homotopy classes of maps $S^1 \to S^1$. This section is devoted to that calculation.

The trick is to open the circle out, and consider maps $[0,1] \to S^1$ instead of $S^1 \to S^1$. Since S^1 can be obtained as a quotient of $[0,1]$ by gluing the endpoints together (see Example 5.51), there is a continuous surjection $\pi : [0,1] \to S^1$, and we will study maps $f : S^1 \to S^1$ by looking at the composites $f \circ \pi : [0,1] \to S^1$.

Having opened the circle out in this way, it turns out that we can "lift" any map $[0,1] \to S^1$ to a map $[0,1] \to \mathbf{R}$ which, when we compose with the exponential map $e : \mathbf{R} \to S^1$ of Example 3.47, gives back the original map $[0,1] \to S^1$. More precisely:

Proposition 6.23 (Path Lifting)

If $g : [0,1] \to S^1$ is a continuous function and $x \in \mathbf{R}$ is any point such that $e(x) = g(0)$, then there is a unique continuous function $\tilde{g} : [0,1] \to \mathbf{R}$ such that $e\tilde{g}(t) = g(t)$ for all $t \in [0,1]$ and $\tilde{g}(0) = x$. So the following triangle commutes:

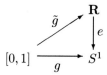

One way to think of this statement is to consider the parameter $t \in [0,1]$ as specifying a moment in time. As t runs from 0 to 1, so g traces out a path in S^1. The condition $e\tilde{g}(t) = g(t)$ specifies that $\tilde{g}(t)$ must always be above $g(t)$ in the spiral picture of Example 3.47. It is as if one person is walking around a circle, and someone else is on a spiral staircase and determined always to be directly above the first person. Clearly they can always do that if they move fast enough (this is the existence part of the proposition), but there is no choice about where they move (this is the uniqueness part).

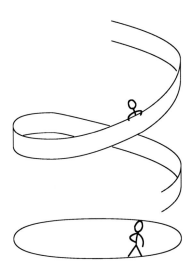

Proof

We will construct $\tilde{g}$ bit by bit. The key to this is that if we take any **proper subset** U of S^1 (i.e., any subset other than the whole of S^1), then its preimage under e is a disjoint union of infinitely many spaces, each homeomorphic to U. Now suppose we have a small interval $[\delta_1, \delta_2] \subset [0,1]$ whose image, under g, is contained in U. And suppose that $\tilde{g}(\delta_1)$ is already defined in such a way that $e\tilde{g}(\delta_1) = g(\delta_1)$. Then $\tilde{g}(\delta_1)$ lies in one of these spaces homeomorphic to U. We can then compose that homeomorphism with g to define $\tilde{g}$ on the interval $[\delta_1, \delta_2]$ so as to agree with the value on δ_1.

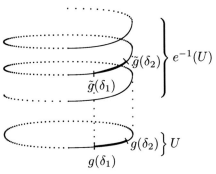

If we can split the interval $[0,1]$ into a number of sections $[\delta_i, \delta_{i+1}]$, with $1 \le i \le n$, such that each section is mapped into some proper subset of S^1, then we can define $\tilde{g}$ inductively over the whole of $[0,1]$, working with one of these sections at a time.

It is enough just to use two subsets of S^1, and we will use $U = S^1 - \{(1,0)\}$ and $V = S^1 - \{(-1,0)\}$. So both U and V are proper subsets and, between them, they contain every point of S^1. Note also that U and V are open, so the preimages $g^{-1}(U)$, $g^{-1}(V)$ will be open sets whose union is $[0,1]$. We can write $g^{-1}(U)$ and $g^{-1}(V)$ as a union of basic open sets, i.e., intervals (a,b), $[0,b)$ or $(a,1]$. This gives an open cover of $[0,1]$ where each set in the cover is a basic open set, and maps into either U or V. Since $[0,1]$ is compact, we can take a finite refinement of this cover, to get a list $I_1, \ldots, I_n$. For convenience we will assume that this is as small a cover as possible, i.e., we cannot delete any of the I_i and still have a cover.[1]

Let us agree to order these open sets in the following way. First, 0 is contained in one of these sets; let that set be I_1. Then $I_1 = [0, b_1)$ for some $0 < b_1 \leq 1$, and b_1 must be contained in another of these sets; let that set be I_2. Then $I_2 = (a_2, b_2)$ where $a_2 < b_1 < b_2$, or $I_2 = (a_2, 1]$. In the first case, b_2 must be contained in another set; let that set be I_3. In the second case, I_1 and I_2 cover $[0,1]$, so we can take $n = 2$. And so forth.

In other words, we put the sets $I_1, \ldots, I_n$ in the order in which we meet them as we travel from 0 to 1.

Let $\delta_0 = 0$, $\delta_n = 1$ and, for $1 \leq i \leq n-1$, $\delta_i = (a_{i+1} + b_i)/2$. Then $\delta_i \in I_i \cap I_{i+1}$ for $1 \leq i \leq n-1$ and so $[\delta_i, \delta_{i+1}] \subset I_{i+1}$ for $0 \leq i \leq n-1$.

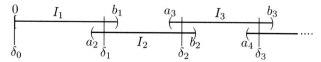

We must have $\tilde{g}(0) = x$, so $\tilde{g}(\delta_0) = x$. Now $[\delta_0, \delta_1] \subset I_1$, and $g(I_1)$ is either contained in U or contained in V. In either case, there is a unique open interval of $\mathbf{R}$, containing x, and homeomorphic with U or V, whichever contains $g(I_1)$. We compose such a homeomorphism with g to get a continuous map $\tilde{g} : I_1 \to \mathbf{R}$ which sends δ_0 to x and is such that $e \circ \tilde{g} = g|I_1$.

In particular, we have defined $\tilde{g}(\delta_1)$. We can use the same argument again to define $\tilde{g}$ on the interval $[\delta_1, \delta_2]$, agreeing with the definition of $\tilde{g}(\delta_1)$. By the gluing lemma, 5.73, the extension of $\tilde{g}$ over $[0, \delta_2]$ is continuous.

Carrying on in the same way, we can define $\tilde{g}$ on the whole of $[0,1]$ and we have a continuous map $\tilde{g} : [0,1] \to \mathbf{R}$ such that $\tilde{g}(0) = x$ and $e \circ \tilde{g} = g$.

Finally, we must prove that the lifting $\tilde{g}$ is unique. So suppose that $\bar{g} : [0,1] \to \mathbf{R}$ is another lift of g with $\bar{g}(0) = x = \tilde{g}(0)$. Since $e \circ \bar{g} = e \circ g$ we see that $\bar{g}(y) - \tilde{g}(y) \in \mathbf{Z}$ for all y. Thus we get a continuous map $\bar{g} - \tilde{g} : [0,1] \to \mathbf{Z}$. By Lemma 4.18, this map must be constant. Since $\bar{g}(0) = \tilde{g}(0) = x$, we conclude that $\bar{g}(y) - \tilde{g}(y) = 0$ for all y, i.e., $\bar{g} = \tilde{g}$. Hence the lift $\tilde{g}$ is unique. □

---

[1] The enthusiastic reader is encouraged to consider what changes to the proof would be necessary if this assumption were removed!

If we take a continuous map $f : S^1 \to S^1$ and form the composite $g = f \circ \pi : [0,1] \to S^1$, then $g(0) = g(1)$. So if we apply this proposition to g, the resulting lift $\tilde{g}$ satisfies $e\tilde{g}(0) = e\tilde{g}(1)$. Now $e(t) = e(s)$ if, and only if, $t - s$ is an integer. So $\tilde{g}(1) - \tilde{g}(0) \in \mathbf{Z}$. Thus, for each map $f : S^1 \to S^1$, we obtain an integer $\tilde{g}(1) - \tilde{g}(0)$, which we call the **degree**, or **winding number**, of f, written $\deg(f)$. Of course, we need to verify that this only depends on f and not on the choice of lifting $\tilde{g}$. But by the uniqueness condition, we know that $\tilde{g}$ is determined by its start point $\tilde{g}(0)$. And this must be such that $e\tilde{g}(0) = g(0)$. Hence, if $\tilde{g}$, $\bar{g}$ are two lifts such that $e\tilde{g} = e\bar{g}$, then $e\tilde{g}(0) = e\bar{g}(0)$ and, so, $\tilde{g}(0) = \bar{g}(0) + c$ for some integer c. Then $x \mapsto \bar{g}(x) + c$ gives another lift of g which agrees with $\tilde{g}$ at 0. Hence, by the uniqueness, $\tilde{g} = \bar{g} + c$. In particular, $\tilde{g}(1) - \tilde{g}(0) = \bar{g}(1) + c - (\bar{g}(0) + c) = \bar{g}(1) - \bar{g}(0)$. In other words, $\tilde{g}$ and $\bar{g}$ give the same answer for the degree of g. Hence this degree does not depend on the choice of lifting.

Example 6.24

Any constant function $S^1 \to S^1$ has degree 0, for the composite $g = f \circ \pi$ will be constant, and the lift $\tilde{g}$ can be taken to be constant: If $x \in \mathbf{R}$ is such that $e(x) = g(0)$, and we define $\tilde{g}$ by $\tilde{g}(t) = x$ for all t, then $e\tilde{g}(t) = e(x) = g(0)$. Hence $\deg(f) = 0$.

Example 6.25

The identity map $S^1 \to S^1$ has degree 1. For $g : [0,1] \to S^1$ is the map $g(t) = (\cos(2\pi t), \sin(2\pi t))$, and a lift is given by $\tilde{g}(t) = t$.

Example 6.26

If f is the map
$$f(\cos(\theta), \sin(\theta)) = (\cos(2\theta), \sin(2\theta)),$$
so that $g : [0,1] \to S^1$ is the map $t \mapsto (\cos(4\pi t), \sin(4\pi t))$, then a lift $\tilde{g}$ is given by $\tilde{g}(t) = 2t$, so $\deg(f) = 2$.

Example 6.27

If n is an integer and f is the map
$$f(\cos(\theta), \sin(\theta)) = (\cos(n\theta), \sin(n\theta)),$$
so that $g : [0,1] \to S^1$ is the map $t \mapsto (\cos(2n\pi t), \sin(2n\pi t))$, then a lift $\tilde{g}$ is given by $\tilde{f}(t) = nt$, so $\deg(f) = n$.

So, we have constructed an integer, the degree, for any map $S^1 \to S^1$. This has not yet told us anything about homotopy classes of maps $S^1 \to S^1$. However, it turns out that homotopic maps have equal degrees. This is proved using the following variant of Proposition 6.23, which shows that just as we can lift paths, so we can lift homotopies.

Proposition 6.28 (Homotopy Lifting)

If $F : [0,1] \times [0,1] \to S^1$ is a continuous function, and $x \in \mathbf{R}$ is any point such that $e(x) = F(0,0)$, then there is a unique continuous function $\tilde{F} : [0,1] \times [0,1] \to \mathbf{R}$ such that $e\tilde{F}(s,t) = F(s,t)$ for all $s,t \in [0,1]$ and $\tilde{F}(0,0) = x$. So the following triangle commutes:

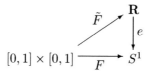

The basic idea of the proof is, as you would expect, the same as for Proposition 6.23. But splitting the square $[0,1] \times [0,1]$ into smaller chunks requires more care than splitting the interval $[0,1]$. Since the ideas we need to split the square will be used a few more times in the book, we present them in a slightly more general form here.

To simplify the statement of the next result, we say that a subset of $\mathbf{R}^n$ has **diameter** less than d if the distance between any pair $\mathbf{x}, \mathbf{y}$ of points in the subset is less than d.

Proposition 6.29 (Domain Splitting)

Suppose we have a map $f : X \to Y$, where X is a compact subset of $\mathbf{R}^n$, and an open cover $\mathcal{U}$ of Y. Then there is some number $\delta > 0$ such that whenever V is a subset of X of diameter less than δ, its image $f(V)$ is contained in one of the sets in $\mathcal{U}$.

Proof

As the map f is continuous, the preimages of the open sets in $\mathcal{U}$ will be open sets in X and these will give an open cover $\mathcal{W}$ of X. Any subset V of X which is contained in one of the sets in $\mathcal{W}$ will, then, have the property that its image, $f(V)$, is contained in one of the sets in $\mathcal{U}$.

The number δ then comes from the Lebesgue lemma, stated next. $\qquad\square$

Lemma 6.30 (Lebesgue Lemma)

Given a compact subspace X of $\mathbf{R}^n$ and an open cover $\mathcal{U}$ of X, there is some $\delta > 0$ such that any subset U of X of diameter less than δ is contained in one of the sets in $\mathcal{U}$.

Proof

Since X is compact, we can refine $\mathcal{U}$ to a finite list $U_1, \ldots, U_n$ of open subsets of X. Then, for $1 \leq i \leq n$, define $f_i : X \to \mathbf{R}$ by setting $f_i(x)$ to be the largest radius r such that $B_r(x)$ is contained in U_i. We take $f_i(x)$ to be 0 if $x \notin U_i$. This is continuous, as is more easily seen by considering f_i as the distance from x to the point in $X - U_i$ nearest to x. Thus the function $f : X \to \mathbf{R}$ defined by $f(x) = \max\{f_i(x) : 1 \leq i \leq n\}$ is also continuous. This function gives the largest radius r such that $B_r(x)$ is contained in one of the open sets U_i.

If there is some $\delta > 0$ such that $f(x) \geq \delta$ for all x, then every open ball of radius less than δ is contained in some open set U_i. Every set of diameter less than δ is contained in an open ball of radius δ, and so the lemma follows.

To see that there is such a δ, note that $f(x) > 0$ for all x, so 0 is not in the image of f. Since X is compact, Proposition 4.27 shows that the image of f will be a compact subset of $\mathbf{R}$. By the Heine–Borel Theorem 4.29, it is thus a closed subset of $\mathbf{R}$, so its complement is open. As this complement contains 0, it also contains some interval $(-\delta, \delta)$ around 0. Hence $f(x) \geq \delta$ for all x. $\qquad\square$

Proof (of Proposition 6.28)

Suppose, then, that we have a homotopy $F : [0,1] \times [0,1] \to S^1$. By covering S^1 with the open sets $U = S^1 - \{(1,0)\}$, $V = S^1 - \{(-1,0)\}$ as before, we can obtain a number $\delta > 0$ such that any subset of $[0,1] \times [0,1]$ of diameter less than δ is mapped into either U or V by F.

We split $[0,1] \times [0,1]$ into an $n \times n$ grid, where n is chosen so that $1/n < \delta/\sqrt{2}$, i.e., each square has diameter less than δ. Hence each square is mapped by F into either U or V.

If $\tilde{F}(0,0) = x$, then that determines a component of $e^{-1}(U)$ or $e^{-1}(V)$ and, hence, a homeomorphism between that component and U or V. By this homeomorphism, we define $\tilde{F}$ on the square $[0, \frac{1}{n}] \times [0, \frac{1}{n}]$. In particular, this defines $\tilde{F}(0, \frac{1}{n})$ and, by the same process, we can define $\tilde{F}$ on the square $[0, \frac{1}{n}] \times [\frac{1}{n}, \frac{2}{n}]$. However, this means defining $\tilde{F}$ on the path $[0, \frac{1}{n}] \times \frac{1}{n}$, based on its value at $(0, \frac{1}{n})$. The problem is that we have already defined $\tilde{F}$ on $[0, \frac{1}{n}] \times \frac{1}{n}$ when we defined it on the square $[0, \frac{1}{n}] \times [0, \frac{1}{n}]$, so we have two definitions which may contradict each other. Fortunately, the uniqueness of path lifting ensures that this cannot happen – if these two paths agree on $(0, \frac{1}{n})$ then they agree everywhere. Hence we can $\tilde{F}$ define on $[0, \frac{1}{n}] \times [0, \frac{2}{n}]$ without problem. Similarly, we can define $\tilde{F}$ on $[0, \frac{1}{n}] \times [0, \frac{3}{n}]$, and so on, until we have $\tilde{F}$ defined on the entire strip $[0, \frac{1}{n}] \times [0, 1]$. Then we use the definition of $\tilde{F}(\frac{1}{n}, 0)$ to define $\tilde{F}$ on the square $[\frac{1}{n}, \frac{2}{n}] \times [0, \frac{1}{n}]$. This entails redefining $\tilde{F}$ on the edge $\frac{1}{n} \times [0, \frac{1}{n}]$ but, again, the uniqueness of path lifting ensures that this definition agrees with the previous one. Next we define $\tilde{F}$ on the square $[\frac{1}{n}, \frac{2}{n}] \times [\frac{1}{n}, \frac{2}{n}]$ based on its value at $(\frac{1}{n}, \frac{1}{n})$. This entails redefining $\tilde{F}$ on two edges $\frac{1}{n} \times [\frac{1}{n}, \frac{2}{n}]$ and $[\frac{1}{n}, \frac{2}{n}] \times \frac{1}{n}$. However, the uniquess of path lifting can be used in both cases to show that the new definition agrees with the old. Then, in a similar way, we can define $\tilde{F}$ on the rest of the strip $[\frac{1}{n}, \frac{2}{n}] \times [0, 1]$ and, continuing similarly, on the whole of the square $[0, 1] \times [0, 1]$.

As before, this lift is unique as, if $\bar{F}$ is a different lift, then $\bar{F}(s, t) - \tilde{F}(s, t)$ is an integer for all $(s, t) \in [0, 1] \times [0, 1]$. As both $\bar{F}$ and $\tilde{F}$ are continuous, this integer must be constant, i.e., independent of s, t. If $\bar{F}(0, 0) = \tilde{F}(0, 0)$, then this integer must be 0, i.e., $\bar{F} = \tilde{F}$. □

Having now established that homotopies can be lifted, we can deduce that homotopic maps have the same degree.

Corollary 6.31

If $f, g : S^1 \to S^1$ are homotopic, then $\deg(f) = \deg(g)$.

Proof

Let $H : S^1 \times [0, 1] \to S^1$ be a homotopy between f and g. Considered as a map defined on $[0, 1] \times [0, 1]$, we can lift this to a map $\tilde{H} : [0, 1] \times [0, 1] \to \mathbf{R}$. Then $\tilde{H}$ restricted to $[0, 1] \times \{0\}$ will give a lift for f, so $\deg(f) = \tilde{H}(1, 0) - \tilde{H}(0, 0)$. And $\tilde{H}$ restricted to $[0, 1] \times \{1\}$ will give a lift for g, so that $\deg(g) = \tilde{H}(1, 1) - \tilde{H}(0, 1)$. In fact, we can use $\tilde{H}$ to define a continuous map $D : [0, 1] \to \mathbf{Z}$ by $D(t) = \tilde{H}(1, t) - \tilde{H}(0, t)$. Then $\deg(f) = D(0)$ and $\deg(g) = D(1)$. However, by Lemma 4.18, such a function D must be constant, since $[0, 1]$ is connected. Hence $\deg(f) = \deg(g)$. □

Corollary 6.32

The circle is not contractible.

Proof

Suppose that $f : S^1 \to \{0\}$ and $g : \{0\} \to S^1$ were homotopy equivalences. So $g \circ f \simeq 1_{S^1}$. Now, $(g \circ f)(x,y) = g(0)$ for all $(x,y) \in S^1$, i.e., this composite is a constant function, and hence has degree 0. Conversely, the identity map has degree 1. As these are different, $g \circ f$ cannot be homotopic to the identity map, so S^1 is not contractible. $\qquad\qquad\square$

To compute $[S^1, S^1]$, we also need the following converse to Corollary 6.31:

Theorem 6.33

If $f, g : S^1 \to S^1$ are such that $\deg(f) = \deg(g)$, then f and g are homotopic.

Proof

The idea is to define a homotopy "upstairs". For simplicity we will first prove this for functions f, g that satisfy $(f \circ \pi)(0) = (g \circ \pi)(0)$. This condition means that we can lift f and g to maps $\tilde{f}, \tilde{g} : [0,1] \to \mathbf{R}$ which satisfy $\tilde{f}(0) = \tilde{g}(0)$. Hence

$$\tilde{f}(1) = \deg(f) + \tilde{f}(0) = \deg(g) + \tilde{g}(0) = \tilde{g}(1).$$

Thus if we define $\tilde{H} : [0,1] \times [0,1] \to \mathbf{R}$ by

$$\tilde{H}(s,t) = t\tilde{f}(s) + (1-t)\tilde{g}(s),$$

then $\tilde{H}(0,t) = \tilde{f}(0) = \tilde{g}(0)$ does not depend on t, and $\tilde{H}(1,t) = \tilde{f}(1) = \tilde{g}(1)$ is, similarly, independent of t. In particular, $\tilde{H}(1,t) - \tilde{H}(0,t) = \deg(f)$ is an integer. Hence when we compose with the exponential map $e : R \to S^1$, we find that $(e \circ \tilde{H})(0,t) = (e \circ \tilde{H})(1,t)$, so we can consider this as a map $H : S^1 \times [0,1] \to S^1$, which is a homotopy between f and g.

If $(f \circ \pi)(0) \neq (g \circ \pi)(0)$, then we use the following lemma to replace g by a function which does agree with f on $\pi(0)$. $\qquad\qquad\square$

Lemma 6.34

If $g : S^1 \to S^1$ and $(x,y) \in S^1$, then there is a map $h : S^1 \to S^1$ which is homotopic to g and such that $h(\pi(0)) = (x,y)$.

Proof

Let θ be the angle from $g(\pi(0))$ to (x,y). Define $H : S^1 \times I \to S^1$ so that $H((x',y'),t)$ is the rotation of (x',y') through the angle $t\theta$. Hence

$H((x', y'), 0) = (x', y')$, and $H((x', y'), 1)$ is (x', y') rotated by θ. In particular, $H(g(\pi(0)), 1) = (x, y)$, while $H(g(x, y), 0) = g(x, y)$. As H is continuous, it gives a homotopy from g to the map $h : S^1 \to S^1$ defined by $h(x', y') = H(g(x', y'), 1)$ which satisfies $h(\pi(0)) = H(g(\pi(0)), 1) = (x, y)$. $\square$

Having completed the proof of Theorem 6.33, we can now give the promised calculation of the set of homotopy classes of self-maps of S^1.

Corollary 6.35

The set of homotopy classes of maps $S^1 \to S^1$ is in one-to-one correspondence with the set of integers, i.e., $[S^1, S^1] = \mathbf{Z}$.

Proof

Every continuous map $S^1 \to S^1$ has a degree, which is an integer. Homotopic maps have the same degree, and non-homotopic maps have different degrees. Hence $[S^1, S^1] \subset \mathbf{Z}$. To complete the proof, we note that all integers occur as the degree of a map, since, for any $n \in \mathbf{Z}$, the map $z \mapsto z^n$ ($z \in \mathbf{C}, |z| = 1$) has degree n, by Example 6.27. $\square$

6.4 Brouwer's Fixed-Point Theorem

We have already seen a theorem saying that any continuous map from $[0, 1]$ to itself must have a fixed point. Our study of continuous maps from S^1 to S^1 can be used to prove a two-dimensional version of this theorem, due to Brouwer.

Theorem 6.36 (Brouwer's Fixed-Point Theorem)

Let $f : D^2 \to D^2$ be a continuous map, where D^2 is the closed disc

$$D^2 = \{(x, y) \in \mathbf{R}^2 : x^2 + y^2 \leq 1\}.$$

Then f has a fixed point, i.e., there is some point $(x, y) \in D^2$ with the property that $f(x, y) = (x, y)$.

Proof

Suppose that $f : D^2 \to D^2$ does not have a fixed point, so that $f(x, y) \neq (x, y)$ for all $(x, y) \in D^2$. So, for each point $(x, y) \in D^2$ we get two points (x, y) and

$f(x, y)$, and we can draw a line through them both. Extend this line beyond (x, y) until it meets the boundary of D^2 (i.e., S^1), and let $g(x, y)$ be the point where this happens. So we get a function $g : D^2 \to S^1$ as in the picture.

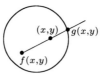

This map g is continuous, essentially because if (x', y') is sufficiently close to (x, y), then $f(x', y')$ will be close to $f(x, y)$ (since f is continuous) and, hence, $g(x', y')$ will be reasonably close to $g(x, y)$. More rigorously, if A is an open arc around $g(x, y)$, then there is some radius r such that whenever (x', y') is in the open ball $B_r(x, y)$ and $f(x', y')$ is in the open ball $B_r(f(x, y))$, then $g(x', y')$ is in A, as depicted below, where A is indicated by a bold line, and the balls around (x, y) and $f(x, y)$ are indicated by the dotted circles of their perimeters. Any straight line which passes through both balls will hit the circle in the region A.

Since f is continuous, there is some radius δ such that $f(x', y') \in B_r(f(x, y))$ whenever $(x', y') \in B_\delta(x, y)$. Hence the preimage $g^{-1}(A)$ contains $B_{\min(\delta, r)}(x, y)$. The same argument can be applied to any point in the preimage, so $g^{-1}(A)$ is open, i.e., g is continuous.

If (x, y) is on the boundary of D^2, then $g(x, y) = (x, y)$ no matter what $f(x, y)$ is.

Now define a map

$$F : S^1 \times I \to S^1$$

by $F((x, y), t) = g(tx, ty)$.

This map F is continuous, so we can think of it as a homotopy between the map $h : S^1 \to S^1$ defined by $h(x, y) = F((x, y), 0)$ and $j : S^1 \to S^1$ defined by $j(x, y) = F((x, y), 1)$. Now $h(x, y) = g(0, 0)$ for all (x, y), so h is the constant map and thus $\deg(h) = 0$. On the other hand, however, $j(x, y) = g(x, y) = (x, y)$ for all (x, y), so j is the identity map and $\deg(j) = 1$. If F is a homotopy between h and j, then these degrees must be equal. Since they are not, the map F cannot exist. Hence nor can g, showing in turn that the map f must have had a fixed point in the first place. $\square$

6.5 Vector Fields

One of the most celebrated theorems of topology is the "Hairy ball theorem". In simple language this says that you cannot comb a hairy ball. To make this more precise, we need the notion of a "vector field".

When combing a surface, such as the sphere, we move the comb in a certain direction, tangential to the surface. This gives a function which assigns, to each point on the surface, a direction, i.e., a vector which is tangential to the surface. For example, if we comb the sphere S^2, we will get a function $v : S^2 \to \mathbf{R}^3$ with the property that $v(s)$ is tangential to the surface of S^2 at s.

In general, combing a surface $S \subset \mathbf{R}^n$ will give rise to a function $v : S \to \mathbf{R}^n$. Of course, this function v should be continuous, as the comb is presumed to move in a continuous way. A continuous tangential vector-valued function such as this is called a **vector field**.

Example 6.37

At any given moment in time there is a vector field which assigns to each point on the surface of the earth, a vector representing the wind felt at that point.

Example 6.38

Another example of a vector field is given by combing a hairy cylinder.

If we constantly comb round the cylinder, then we get a nowhere-zero vector field, i.e., $v(s) \neq 0$ for all s in the cylinder.

We can use our knowledge about homotopy classes of maps from S^1 to S^1 to tell us about vector fields, as illustrated by the next two theorems.

Theorem 6.39

If you stir a cup of coffee, then, at any given moment in time, some particle on the surface is stationary.

Proof

Let $v : D^2 \to \mathbf{R}^2$ be the vector field indicating how the surface of the coffee is moving, so that $v(x, y)$ is the velocity of a particle of coffee at the point (x, y). If v is nowhere zero, then we can define a continuous map $g : S^1 \to S^1$ by

$$g(x, y) = \frac{-v(x, y)}{|v(x, y)|},$$

thinking of $(x, y) \in S^1$ as a point on the boundary of D^2. Then there is a homotopy $G : g \simeq id$ defined by

$$G((x, y), t) = \frac{t(x, y) - (1 - t)v(x, y)}{|t(x, y) - (1 - t)v(x, y)|}.$$

It takes some thought to see that this is continuous, as we must verify that the denominator $|t(x, y) - (1 - t)v(x, y)|$ cannot be zero. If it were zero, then this would say that $t(x, y) = (1 - t)v(x, y)$. If $t = 0$, then that would mean $v(x, y) = 0$, which cannot happen by assumption. If $t = 1$, then that would mean that $(x, y) = 0$, which cannot happen as $(x, y) \in S^1$. If $0 < t < 1$, then $t(x, y) - (1 - t)v(x, y) = 0$ implies that $v(x, y) = \frac{t}{1-t}(x, y)$, i.e., $v(x, y)$ is a positive multiple of (x, y). Since (x, y) is on the perimeter of the cup, this would be saying that the coffee is moving out of the cup, which cannot happen. Hence $|t(x, y) - (1 - t)v(x, y)|$ is never zero, so G is a continuous map.

On the other hand, if v is nowhere zero, then we can also define a homotopy $F : S^1 \times I \to S^1$ by

$$F((x, y), t) = \frac{-v(tx, ty)}{|v(tx, ty)|}.$$

If $t = 1$, then $F((x, y), 1) = g(x, y)$ and if $t = 0$, $F((x, y), 0) = -v(0, 0)/|v(0, 0)|$ is constant. So F is a homotopy between g and a constant map. Putting these homotopies together, we get

$$id \simeq g \simeq \text{constant}.$$

A constant map has degree 0, and the identity has degree 1, hence these two cannot be homotopic. So v must be zero somewhere, i.e., some point is stationary. $\qquad\square$

Theorem 6.40 (Hairy Ball Theorem)

Let $v : S^2 \to \mathbf{R}^3$ be a vector field on the sphere. Then there is some point $(x, y, z) \in S^2$ such that $v(x, y, z) = 0$.

Proof

To prove this, we will split the sphere up into three sections, by latitude:

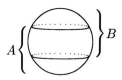

We will first consider the region A, i.e., everything below (and including) the upper line. This region is homeomorphic to the closed disc D^2, by a version of stereographic projection (e.g. the projection defined in Example 5.7 followed by a function that scales the region corresponding to A down to the size of D^2). More importantly, if we think of v as placing an arrow at each point on S^2 tangential to S^2, then under this stereographic projection, v corresponds to a continuous map $\tilde{v}$ from D^2 to $\mathbf{R}^2$, placing an arrow at every point of D^2 tangential to D^2. (This correspondence sounds plausible for points near the South Pole. To prove that it works for the whole region A requires methods from multivariate calculus whose details we omit. See Section 7c of [4] for more information, or Section 2.2 of [5] for a different approach.)

Now we modify $\tilde{v}$ as follows. Let $h : D^2 \to D^2$ be the continuous map which shrinks the disc of radius $1/2$ within D^2 down to a point and stretches out the remainder of D^2 accordingly.

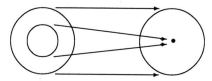

So, using polar coordinates, h can be written as

$$h(r\cos(\theta), r\sin(\theta)) = \begin{cases} ((2r-1)\cos(\theta), (2r-1)\sin(\theta)) & \text{if } r \geq \frac{1}{2}, \\ (0,0) & \text{if } r \leq \frac{1}{2}. \end{cases}$$

We define a new function $\tilde{w} : D^2 \to \mathbf{R}^2$ by $\tilde{w} = \tilde{v} \circ h : D^2 \to \mathbf{R}^2$, and we can think of this as a vector field on D^2. Note that $\tilde{w}(x,y) = \tilde{v}(0,0)$ if $|(x,y)| \leq 1/2$, i.e., $\tilde{w}$ is constant throughout the disc of radius $1/2$ inside D^2. But when $|(x,y)| = 1$, $\tilde{w}(x,y) = \tilde{v}(x,y)$, so $\tilde{v}$ and $\tilde{w}$ agree with each other on the perimeter of D^2.

The fact that $\tilde{v}$ and $\tilde{w}$ agree on the perimeter of D^2 means that we can patch $\tilde{w}$ in, in place of $\tilde{v}$, in our original vector field on S^2, to get a new continuous tangential vector field w. Because $\tilde{w}$ is constant in the middle of D^2, so w is "constant" throughout a disc-like region around the south pole. We assume that the lower tropic, i.e., the southern boundary of B, is precisely the boundary of this 'constant' region. So the modified vector field, w, is constant on the boundary and the complement of B.

The region B itself is, by another version of stereographic projection, also homeomorphic to D^2. But look what happens to the perimeter of this region under this homeomorphism. On the sphere, this perimeter corresponds to the lower tropic, on which w is constant, depicted in the left of the following picture.

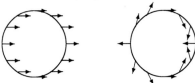

However, when we apply stereographic projection, it has the effect of turning this circle inside out. The arrows on the perimeter then point in different directions. In fact, as you can see from the right-hand picture above, the arrows rotate through 720° as you pass around the perimeter circle.

We can use this to define a continuous map $f : S^1 \to S^1$ by

$$f(x, y) = \frac{w'(x, y)}{|w'(x, y)|},$$

where $w' : D^2 \to \mathbf{R}^2$ is the function corresponding to w under stereographic projection. Since the arrows rotate through 720° as we go around the circle, so $\deg(f) = 2$. However, if v has no zeros, then w will have no zeros and so, in particular, w' will be nowhere zero. We could then define a homotopy $H : S^1 \times I \to S^1$ by

$$H((x, y), t) = \frac{w'(t(x, y))}{|w'(t(x, y))|}.$$

If w' is nowhere zero, then this is continuous. When $t = 0$, $H((x, y), 0) = w'(0, 0)/|w'(0, 0)|$ is constant. So H would be a homotopy from a degree 2 map to a constant map. Since this cannot happen, w' must have a zero somewhere. Consequently, so must w and, in turn, v. □

Since, as we have seen, wind can be considered as a vector field on the surface of the earth, which is homeomorphic to S^2, we get the following meteorological consequence.

Corollary 6.41

At any moment in time, there is some point on the earth where there is no wind.

Remark: Notice how we have applied our knowledge of $[S^1, S^1]$ to problems about vector fields, which don't directly involve S^1 or homotopy. A little knowledge can, indeed, go a long way, and this shows how useful the "homotopy" concept is that it can solve problems that have no apparent connection to it.

EXERCISES

6.1. Write down a homotopy equivalence between $(0, 1)$ and $[0, 1]$.

6.2. List all homotopy classes of maps $(0, 1) \to (0, 1)$.

6.3. Prove that a discrete space consisting of m points is homotopy equivalent to a discrete space consisting of n points if, and only if, $m = n$.

6.4. Let X be any space and $f : X \to S^n$ a continuous map. Using Proposition 6.5, show that if f is not surjective, then f is homotopic to a constant map.

6.5. Show that the map $f : S^1 \to S^1$ given by $f(x, y) = (-x, -y)$ is homotopic to the identity map.

6.6. If $f, g : S^1 \to S^1$ are two continuous maps, express $\deg(f \circ g)$ in terms of $\deg(f)$ and $\deg(g)$. Use this to show that $f \circ g$ is homotopic to $g \circ f$.

6.7. Which of the following surfaces do you think can be combed (i.e., which admit a nowhere-zero tangential vector field): (1) a Möbius band, (2) a surface of genus two, (3) a torus, and (4) a Klein bottle?

7
The Euler Number

We now begin our study of topological invariants, by considering the "Euler number" or "Euler characteristic." This assigns an integer to each topological space in a way that tells us something about the topology of the space. In particular, it can sometimes tell if two spaces are not homotopy equivalent, since spaces which are homotopy equivalent have the same Euler number.

7.1 Simplicial Complexes

Although it is possible to define the Euler number for all spaces, for clarity we will begin by restricting our attention to "simplicial complexes." These are spaces built out of cells, called simplices.[1] For example, here are simplicial complexes homeomorphic to the circle, the solid square, and the annulus.

These have been built out of points, lines and triangles, i.e., 0-simplices, 1-simplices and 2-simplices. Essentially, a k-simplex is described by a list of $k+1$

---

[1] For reasons which I cannot fathom, the plural of "simplex" is "simplices" whereas the plural of "complex" is "complexes"!

M.D. Crossley, *Essential Topology*, Springer Undergraduate
Mathematics Series, DOI 10.1007/978-1-84628-194-5_7,
© Springer-Verlag London Limited 2010

vertices (points in some $\mathbf{R}^n$), and is the smallest convex subspace of $\mathbf{R}^n$ containing those vertices. Hence, for example, a 1-simplex is defined by 2 vertices and is the straight line from one of these vertices to the other.

However, we need to be careful about our choice of vertices in order to avoid degenerate situations such as the following:

These both have 4 vertices, so would be 3-simplices by the above definition, whereas neither is 3-dimensional.

To avoid such cases, we insist that the vertices $v_0, \ldots, v_k$ be in **general position**, by which we mean that the k vectors

$$v_1 - v_0, \quad v_2 - v_1, \quad \ldots, \quad v_k - v_{k-1}$$

in $\mathbf{R}^n$ are linearly independent. This ensures that a k-simplex really is k-dimensional, i.e., is not contained in any $(k-1)$-dimensional subspace of $\mathbf{R}^n$.

Thus we define a k-**simplex** as the smallest convex subspace of $\mathbf{R}^n$ containing a given list of $k+1$ vertices which are in general position. We will write $[v_0, \ldots, v_k]$ for the k-simplex with vertices $v_0, \ldots, v_k$, so that $[v_0, \ldots, v_k]$ consists of all linear combinations

$$t_0 v_0 + \cdots + t_k v_k$$

where the coefficients $t_0, \ldots, t_k$ are real numbers between 0 and 1 satisfying $t_0 + \cdots + t_k = 1$ and are called the **barycentric coordinates** of the point $t_0 v_0 + \cdots + t_k v_k$.

If we have a k-simplex with vertices $v_0, \ldots, v_k$, then any non-empty subset of these vertices will also determine a simplex, called a **subsimplex** of the original k-simplex. We will say that a subsimplex is a **face**[2] if it only omits one vertex, so a k-simplex will have $k+1$ faces and $2^{k+1}-1$ subsimplices. For example, a 2-simplex $[v_0, v_1, v_2]$ has 7 subsimplices: $[v_0], [v_1], [v_2], [v_0, v_1], [v_0, v_2], [v_1, v_2]$ and $[v_0, v_1, v_2]$, of which 3 are faces: $[v_0, v_1], [v_0, v_2], [v_1, v_2]$. The union of the faces is called the **boundary** of the simplex. The complement of the boundary is called the **interior** of the simplex. The boundary consists of all points with at least one barycentric coordinate equal to 0, and the interior consists of all points whose barycentric coordinates are *all* non-zero. For 0-simplices (i.e., points), the boundary is actually empty, since a 0-simplex has no proper subsimplices, so the interior is the simplex itself in this case.

---

[2] Some authors use the term "face" to mean subsimplex, but we reserve the word "face" for a $(k-1)$-subsimplex of a k-simplex

A simplicial complex is, essentially, just a finite union of simplices. However, to avoid some technical problems later, we will insist on two extra conditions.

Definition: A **simplicial complex** K is a subspace of $\mathbf{R}^n$ together with a finite list of simplices such that:

1. The union of the simplices is the set K and each point in K lies in the interior of only one simplex.

2. Every face of every simplex in the list is also in the list.

Note that some books allow a simplicial complex to have infinitely many simplices. For most examples, a finite number of simplices is enough and ensures that every simplicial complex is compact, being a finite union of compact sets.

We say that a simplicial complex is n-dimensional if it has at least one n-simplex, but no $(n+1)$-simplices, $(n+2)$-simplices, etc.

Example 7.1

The simplicial circle above is a one-dimensional simplicial complex with three 0-simplices (the vertices of the triangle) and three 1-simplices (the edges of the triangle).

It is tempting to think that one can make a simplicial circle with only two vertices and two curved edges. However, since simplices must be convex, we cannot have curved 1-simplices. Hence we must have at least three vertices in a simplicial circle.

Example 7.2

The simplicial square above is a two-dimensional simplicial complex with four 0-simplices, five 1-simplices and two 2-simplices.

Example 7.3

The simplicial annulus above is a two-dimensional simplicial complex with 6 0-simplices, 12 1-simplices and 6 2-simplices.

With a little work, we can see that the conditions in our definition of simplicial complex ensure that non-empty intersections of simplices are always simplices.

Proposition 7.4

If S and T are simplices of a simplicial complex K, then $S \cap T$ is either empty or a subsimplex of both S and T.

Proof

If $S \cap T$ is not empty, let $v_1, \ldots, v_n$ be the set of all vertices of K that are contained in $S \cap T$. We will prove that $S \cap T$ is a subsimplex of both S and T by showing that $S \cap T$ is the simplex $[v_1, \ldots, v_n]$.

To do this, let $x \in S \cap T$ be any point. By condition 1 of the definition of simplicial complex, x is contained in the interior of exactly one simplex of K. Let $[w_1, \ldots, w_k]$ be that simplex. Since $x \in S$ we can write x as a linear combination of the vertices of S, with non-negative real coefficients that sum to 1. By taking only those vertices of S that have non-zero coefficients in this expression, we can find a subsimplex of S whose interior contains x. A subsimplex of S is a simplex of K by applying condition 2 repeatedly, so this subsimplex must be $[w_1, \ldots, w_k]$ by the uniqueness in condition 1. Hence $w_1, \ldots, w_k$ are vertices in S. The same argument can be applied with T in place of S, from which we see that $w_1, \ldots, w_k \in S \cap T$. Thus $\{w_1, \ldots, w_k\} \subset \{v_1, \ldots, v_n\}$, so $x \in [w_1, \ldots, w_k] \subset [v_1, \ldots, v_n]$. Consequently $S \cap T \subset [v_1, \ldots, v_n]$. On the other hand, since $v_1, \ldots, v_n \in S$, and S is convex, we have $[v_1, \ldots, v_n] \subset S$. The same argument applies to T, revealing that $[v_1, \ldots, v_n] \subset S \cap T$. Hence these two sets coincide. $\square$

7.2 The Euler Number

If we have a simplicial complex, we associate a number to it, the "Euler number" in the following way.

Definition: If T is an n-dimensional simplicial complex and, for each k, i_k is the number of k-simplices in T, then the **Euler number** of T, written $\chi(T)$, is given by

$$\chi(T) = i_0 - i_1 + i_2 - i_3 + \cdots + (-1)^n i_n.$$

In many books the Euler number is called the **Euler characteristic**; we will use the two terms interchangeably.

Example 7.5

The simplicial circle with three 0-cells and three 1-cells has Euler number $\chi = 3 - 3 = 0$.

Example 7.6

The simplicial square with four 0-cells, five 1-cells and two 2-cells has Euler number $\chi = 4 - 5 + 2 = 1$.

Example 7.7

The simplicial annulus with 6 0-cells, 12 1-cells and 6 2-cells has Euler number $\chi = 6 - 12 + 6 = 0$.

Example 7.8

There is a simplicial torus which looks like this:

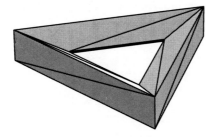

This has 9 0-simplices, 27 1-simplices and 18 2-simplices, so its Euler number is $\chi = 9 - 27 + 18 = 0$.

Having defined the Euler characteristic of a simplicial complex, we can extend this to topological spaces by an appropriate homeomorphism, called a "triangulation". A **triangulation** of a topological space T is a simplicial complex K and a homeomorphism $K \leftrightarrow T$. A space for which such a triangulation exists is said to be **triangulable**.

Example 7.9

One triangulation of the circle S^1 is given by the simplicial complex in $\mathbf{R}^2$ which has three 0-simplices $(0, 2)$, $(\sqrt{3}, -1)$, $(-\sqrt{3}, -1)$ and three 1-simplices between them, together with a homeomorphism between this and S^1, such as $(x, y) \longrightarrow (x, y)/\sqrt{x^2 + y^2}$.

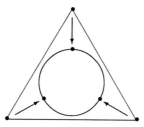

This simplicial complex has Euler number $3 - 3 = 0$.

Example 7.10

Another triangulation of S^1 can be given by taking the simplicial complex K in $\mathbf{R}^2$ that has four 0-simplices $(2,0)$, $(-2,0)$, $(0,2)$, $(0,-2)$ and four 1-simplices between them, with a homeomorphism, such as $(x,y) \mapsto (x,y)/\sqrt{x^2 + y^2}$.

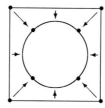

This simplicial complex has Euler number $4 - 4 = 0$.

Example 7.11

We can triangulate the 2-sphere S^2 as a tetrahedron, using four 2-simplices, six 1-simplices and four 0-simplices.

This simplicial complex has Euler number $4 - 6 + 4 = 2$.

Notice that the two different triangulations of S^1 both have the same Euler number. Indeed, it is easy to convince yourself that *any* triangulation of S^1 will have Euler number 0. In fact, the analogous result is true for any triangulable space: Any two different triangulations of the same space will have the same Euler number, although this is very difficult to prove. (One way to prove it is to use Lemma 6.11 and Theorem 7.13 below.) This allows us to define the

Euler number $\chi(T)$ of a triangulable space T to be the Euler number of any simplicial complex K homeomorphic with T. We can then deduce that if two spaces have different Euler numbers, then they cannot be homeomorphic.

Example 7.12

The sphere S^2 has Euler number 2, hence S^2 is not homeomorphic with the torus T^2 which has Euler number 0.

Moreover, we can deduce that two triangulable spaces with different Euler numbers cannot even be homotopy equivalent, thanks to the following result.

Theorem 7.13

If two triangulable spaces are homotopy equivalent, then they have the same Euler number.

This is also very difficult to prove, of course, and we will derive it as a consequence of another very deep theorem at the end of Chapter 10.

Example 7.14

The sphere S^2 has Euler number 2, hence S^2 is not homotopy equivalent with the torus T^2 which has Euler number 0. Moreover, neither S^2 nor T^2 is contractible, since a one-point space has Euler number 1.

7.3 The Euler Characteristic and Surfaces

We have seen that two homeomorphic spaces have the same Euler number. However, many pairs of spaces which are not homeomorphic, or even homotopy equivalent, have the same Euler number. For example, $\chi(S^1) = 0 = \chi(T^2)$, yet the circle and the torus are not homotopy equivalent, as we will see in Example 8.11.

However, for certain types of spaces, the Euler number can distinguish non-homeomorphic spaces. The most important example of this is the classification of triangulable "surfaces." A **surface** is defined to be a Hausdorff space with the property that around every point in the space, there is an open neighbourhood homeomorphic with an open disc in $\mathbf{R}^2$.

Example 7.15

The plane $\mathbf{R}^2$ is a surface, since every point (x, y) is contained in a neighbourhood, say $B_1(x, y)$, which is an open disc in $\mathbf{R}^2$.

Example 7.16

The torus T^2 is a surface. For example, the picture below depicts a disc-like neighbourhood about the point $(3/\sqrt{2}, -3/\sqrt{2}, 0)$.

Example 7.17

Similarly, the sphere S^2 is a surface, as are the genus 2 surface of Example 3.27 and the Klein bottle of Example 5.57.

Example 7.18

The open cylinder $S^1 \times (0, 1)$ is a surface similarly. And if we form a Möbius band out of this by cutting, twisting and gluing, then we will get another surface.

Now, a surface may be orientable or non-orientable. To understand the difference, imagine holding a small sheet of paper against the surface at one point, with an asymmetric figure, such as a spiral, drawn on the paper.

We can slide the paper along the surface and rotate it around and, provided that we do not move very far, the spiral will look the same. However, on some surfaces it is possible to slide the paper around the surface in a certain way and arrive back at the starting point with the spiral reflected. For example, on a Möbius band, if you slide the paper once around the band, this will happen.

In fact, if you do this with a real Möbius band and piece of paper, the paper will end up on the opposite side of the band to where you first started, but we want to imagine this sliding process taking place *inside* the surface, so we will suppose that the surface and the paper are transparent so that we cannot tell which side of the surface the paper is on.

We then say that the Möbius band is **non-orientable** because the orientation of the spiral does not stay constant as you slide around the surface. By contrast, a cylinder is **orientable** as, no matter how you move around the surface, the spiral will have the same orientation.

With this notion of orientability, we have the following amazing result.

Theorem 7.19 (Classification of Surfaces)

Two triangulable surfaces S and T are homeomorphic if, and only if, they have the same Euler number and the same orientability (i.e., either both are orientable or both are non-orientable).

In other words, if we know whether the surfaces are orientable or not, then their Euler number is enough to tell whether the surfaces are homeomorphic to each other or not.

As we have defined things, the condition that S and T be triangulable is necessary for their Euler numbers to be defined. However, it also hides an important hypothesis, which is that S and T be compact for, as we mentioned earlier, every simplicial complex is compact. The theorem, consequently, does not apply to the spaces of Examples 7.15 and 7.18. Some people use the term **closed surface** to mean a compact surface, and so will talk of this theorem as giving a classification of closed surfaces.

The proof of Theorem 7.19 is too long for this book, but it is explained very well in Chapter 1 of [7] and Chapter 17 of [4].

EXERCISES

7.1. Show that every triangulable space is Hausdorff.

7.2. Show directly that any two triangulations of the circle S^1 have the same Euler number. Which other spaces can you give such a direct proof for?

7.3. Using any triangulation that you can think of, calculate the Euler number of (1) a closed interval $[a, b]$, (2) a cylinder, (3) a Möbius band, (4) a surface of genus two, (5) the Klein bottle.

7.4. For each positive integer n, find a simplicial complex with Euler number n. For each positive integer n, try to find a connected simplicial complex with Euler number n.

7.5. Which integers (positive or negative) can occur as the Euler number of a one-dimensional simplicial complex? Which integers can occur as the Euler number of a connected one-dimensional simplicial complex?

7.6. Give an example of a 'non-Hausdorff surface', i.e., a topological space S which is not Hausdorff, but which has the property that every point has an open neighbourhood homeomorphic with an open disc in $\mathbf{R}^2$.

8

Homotopy Groups

In Chapter 6 we calculated the set $[S^1, S^1]$ of homotopy classes of maps $S^1 \to S^1$ and found that $[S^1, S^1] = \mathbf{Z}$, which is an Abelian group. We can describe the group operation topologically as follows.

Imagine taking S^1 and gluing the points $(1,0)$ and $(-1,0)$ together:

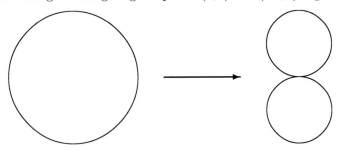

This leaves a space homeomorphic to a pair of circles joined at one point. Now if we have any two continuous maps $f, g : S^1 \to X$, where the range X can be any topological space, then we can define a map $f \# g : S^1 \to X$, by first mapping S^1 to this pair of circles, and then mapping the top circle by f and the bottom by g. However, this may give conflicting definitions on the middle point shared between the two circles. So we need f and g to agree at one point. For convenience we choose that point to be $(1,0)$ in S^1, and we orient the two smaller circles so that their common point is $(1,0)$. So, then, any two continuous maps $f, g : S^1 \to X$ such that $f(1,0) = g(1,0)$ combine to give a new map $f \# g : S^1 \to X$ which is continuous by the gluing lemma 5.73. The map $f \# g$ can be depicted as follows, where a sample of points on the circle has

M.D. Crossley, *Essential Topology*, Springer Undergraduate
Mathematics Series, DOI 10.1007/978-1-84628-194-5_8,
© Springer-Verlag London Limited 2010

been labelled with their image under $f \# g$:

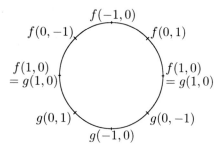

Note that $(f \# g)(1,0)$ is the same as $f(1,0) = g(1,0)$.

If f, g are maps $S^1 \to S^1$, then the degree of $f \# g$ can be seen to equal $\deg(f) + \deg(g)$, as follows. Suppose we have lifts $\tilde{f}$ and $\tilde{g}$ of f and g. Since $g(1,0) = f(1,0)$ and $(1,0) = e(0)$, we can arrange that $\tilde{g}(0) = \tilde{f}(1)$ by adding an integer to $\tilde{g}$ if necessary. Then we can define a lift of $f \# g$ by

$$\widetilde{f \# g}(t) = \begin{cases} \tilde{f}(2t) & \text{if } t \leq \frac{1}{2}, \\ \tilde{g}(2t - 1) & \text{if } t \geq \frac{1}{2}, \end{cases}$$

so that

$$\begin{aligned} \widetilde{f \# g}(1) - \widetilde{f \# g}(0) &= \tilde{g}(1) - \tilde{f}(0) \\ &= \tilde{g}(1) - \tilde{g}(0) + \tilde{f}(1) - \tilde{f}(0) \\ &= \deg(g) + \deg(f). \end{aligned}$$

Hence $\deg(f \# g) = \deg(f) + \deg(g)$, so the operation $\#$ corresponds to addition in $\mathbf{Z}$, and we think of $\#$ as an addition operation in the set of maps $S^1 \to S^1$.

8.1 Homotopy Groups

So far we have an "addition" operation on the set of maps $S^1 \to X$, for any space X, and we know that when we take homotopy classes, this gives an Abelian group, $\mathbf{Z}$, in the case where $X = S^1$. In fact, with a little care, we can always get a group after taking homotopy classes, though not always an Abelian group.

In order to do this neatly, we should take care of that irritating condition that $f(1,0)$ should equal $g(1,0)$. The best way of tidying this up is to work with "pointed spaces." A **pointed space** is a topological space X together with a specific choice of point $x_0 \in X$, called the **base point**. If we wish to emphasize the base point, then we write (X, x_0) for the pointed space X with base point x_0. We then only consider **pointed maps** $f : (X, x_0) \to (Y, y_0)$, i.e., continuous maps which satisfy $f(x_0) = y_0$, and **pointed homotopies** $F : (X, x_0) \times [0, 1] \to (Y, y_0)$, i.e., those which satisfy $F(x_0, t) = y_0$ for all $t \in [0, 1]$.

With the circle S^1, we choose $(1, 0)$ to be the base point, and then any pointed maps $f, g : S^1 \to (X, x_0)$ will satisfy $f(1, 0) = g(1, 0)$, because $f(1, 0) = x_0 = g(1, 0)$. And, as we have noted, the sum $f \# g$ also satisfies $(f \# g)(1, 0) = x_0$, so this is another pointed map.

Hence the set of pointed maps $S^1 \to X$ can be given an addition operation, for any pointed space X. In fact, we can do the same for any sphere S^n, with $n > 0$, by collapsing the "equator" to a single point, i.e., we glue together all the points whose last coordinate is zero.

To make this precise, and to simplify the following discussion about maps from S^n, we will use a bit of deconstructionism on S^n. Recall from Example 5.51 that S^n is homeomorphic to the quotient of the n-dimensional square $[0, 1]^n$ by its boundary, i.e., the set of points with at least one coordinate equal to 0 or 1. Since all the boundary points of $[0, 1]^n$ get glued together to one point, this gives us a natural choice of base point. Then a pointed map $S^n \to X$ corresponds to a continuous map $[0, 1]^n \to X$ that sends all boundary points of $[0, 1]^n$ to the base point of X.

To deal with situations like this, it is convenient to work with "topological pairs" rather than individual spaces. A **topological pair** is a pair (X, A) where X is a topological space and A is a subspace of X. A map $f : (X, A) \to (Y, B)$ of topological pairs is a continuous map $f : X \to Y$ such that $f(A) \subset B$. For example, a pointed map $(X, x_0) \to (Y, y_0)$ is a map of pairs $(X, \{x_0\}) \to (Y, \{y_0\})$. With this example in mind, we write (X, x_0) instead of $(X, \{x_0\})$ so as to keep the notation under control. A pointed map $S^n \to X$ then corresponds to a map of pairs $([0, 1]^n, \partial[0, 1]^n) \to (X, x_0)$, where $\partial[0, 1]^n$ is the boundary of $[0, 1]^n$. As this notation can get clumsy, and we will be using this correspondence a lot in this section, we will write S_n for $[0, 1]^n$, so that a pointed map $S^n \to X$ corresponds to a map $(S_n, \partial S_n) \to (X, x_0)$. (If the domain of a map is a topological pair, then it is always to be assumed that the map is a map of topological pairs. We will often omit the verification that $f(A) \subset B$, because it is usually very straightforward.)

Given two such maps $f, g : (S_n, \partial S_n) \to (X, x_0)$, we define $f \# g$ by

$$(f \# g)(s_1, \ldots, s_n) = \begin{cases} f(s_1, \ldots, s_{n-1}, 2s_n) & \text{if } s_n \leq \frac{1}{2}, \\ g(s_1, \ldots, s_{n-1}, 2s_n - 1) & \text{if } s_n \geq \frac{1}{2}, \end{cases}$$

for $(s_1, \ldots, s_n) \in S_n$.

If $s_n = 1/2$, then $(s_1, \ldots, s_{n-1}, 2s_n)$ and $(s_1, \ldots, s_{n-1}, 2s_n - 1)$ are both in ∂S_n, so $f(s_1, \ldots, s_{n-1}, 2s_n) = g(s_1, \ldots, s_{n-1}, 2s_n - 1) = x_0$. So $f \# g$ is a continuous map on S_n by the gluing lemma, 5.73. Moreover, if $(s_1, \ldots, s_n)$ is in ∂S_n, then so is $(s_1, \ldots, s_{n-1}, 2s_n)$ if $s_n \leq 1/2$, and so is $(s_1, \ldots, s_{n-1}, 2s_n - 1)$ if $s_n \geq 1/2$. So $f \# g$ does take ∂S_n to the base point of X and, hence, is a map of pairs $(S_n, \partial S_n) \to (X, x_0)$.

Thus we have an "addition" operation on the set of maps $(S_n, \partial S_n) \to (X, x_0)$ and, hence, an addition operation on the set of pointed maps $S^n \to X$, which we also denote by $\#$. By suppressing all coordinates except s_1 and s_n, the addition on maps from $(S_n, \partial S_n)$ can be depicted in two dimensions as

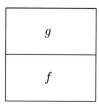

As it stands, this operation is not very well behaved. It is not associative, for example. However, it does respect pointed homotopies, as we will now see.

Proposition 8.1

If $f \simeq f'$ and $g \simeq g'$ by pointed homotopies, then $f \# g \simeq f' \# g'$ by a pointed homotopy.

Proof

If $F : S^n \times [0,1] \to X$ and $G : S^n \times [0,1] \to X$ are pointed homotopies from f to f' and g to g', then a pointed homotopy H from $f \# g$ to $f' \# g'$ can be defined by

$$H_t = F_t \# G_t,$$

where $F_t : S^n \to X$ is the map given by $F_t(\mathbf{x}) \mapsto F(\mathbf{x}, t)$ for $\mathbf{x} \in S^n$, and G_t and H_t are defined similarly.   □

This enables us to derive an addition operation on homotopy classes of maps from S^n. So, if we write $\pi_n(X)$ for the set of pointed homotopy classes of pointed maps $S^n \to X$, then this set has an operation $+$, defined by $[f] + [g] = [f \# g]$, where $f, g : S^n \to X$ are pointed maps, and the square brackets $[f]$ denote the class of all maps which are homotopic to f.

The following three propositions show that $\pi_n(X)$ is actually a group under this operation $+$.

Proposition 8.2

Let $c : S^n \to X$ be the constant map to the base point of X. Then, for any map $f : S^n \to X$, the maps $f\#c$ and $c\#f$ are homotopic to f.

Hence $\pi_n(X)$ has a zero element for the addition operation $+$.

Proof

Given any map $f : S^n \to X$, we will show that $f\#c$ is homotopic to f. As usual, we consider f and c as maps of pairs $(S_n, \partial S_n) \to (X, x_0)$, where x_0 is the base point of X.

We define a homotopy $H : S_n \times [0,1] \to X$ by

$$H((s_1, \ldots, s_n), t) = \begin{cases} x_0 & \text{if } s_n \leq \frac{t}{2}, \\ f(s_1, \ldots, s_{n-1}, \frac{2s_n - t}{2-t}) & \text{if } s_n \geq \frac{t}{2}. \end{cases}$$

Then

$$H((s_1, \ldots, s_n), 1) = \begin{cases} c(s_1, \ldots, s_{n-1}, 2s_n) & \text{if } s_n \leq \frac{1}{2}, \\ f(s_1, \ldots, s_{n-1}, 2s_n - 1) & \text{if } s_n \geq \frac{1}{2} \end{cases}$$
$$= c\#f(s_1, \ldots, s_n)$$

and $H((s_1, \ldots, s_n), 0) = f(s_1, \ldots, s_n)$.

Suppressing all coordinates but s_1 and s_n, we can depict H as

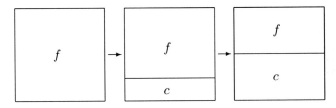

The left-hand picture shows H at $t = 0$, the middle shows H at $t = 1/2$, and the right-hand picture shows H at $t = 1$.

Similarly, $f\#c$ is homotopic to f by a homotopy such as

$$H((s_1, \ldots, s_n), t) = \begin{cases} f(s_1, \ldots, s_{n-1}, \frac{2s_n}{2-t}) & \text{if } s_n \leq 1 - \frac{t}{2}, \\ x_0 & \text{if } s_n \geq 1 - \frac{t}{2}. \end{cases} \qquad \square$$

Proposition 8.3

For each map $f : S^n \to X$, there is another map $\bar{f} : S^n \to X$ such that $f\#\bar{f}$ and $\bar{f}\#f$ are homotopic to the constant map c of Proposition 8.2.

Hence, the addition operation $+$ in $\pi_n(X)$ admits inverses.

Proof

As usual, we work with maps $(S_n, \partial S_n) \to (X, x_0)$, using the same letter as for the map from S^n. Thus, if $f : (S_n, \partial S_n) \to (X, x_0)$, then we define $\bar{f} :$ $(S_n, \partial S_n) \to (X, x_0)$ by

$$\bar{f}(s_1, \ldots, s_n) = f(s_1, \ldots, s_{n-1}, 1 - s_n).$$

A homotopy from $f \# \bar{f}$ to c is given by

$$H((s_1, \ldots, s_n), t) = \begin{cases} (f \# \bar{f})(s_1, \ldots, s_{n-1}, \frac{1-t}{2}) & \text{if } \frac{1-t}{2} \le s_n \le \frac{1+t}{2}, \\ (f \# \bar{f})(s_1, \ldots, s_{n-1}, s_n) & \text{otherwise.} \end{cases}$$

This is clearly consistent at $s_n = (1 - t)/2$ but looks contradictory at $s_n = (1 + t)/2$. However:

$$(f \# \bar{f})(s_1, \ldots, s_{n-1}, \frac{1+t}{2}) = \bar{f}(s_1, \ldots, s_{n-1}, t) = f(s_1, \ldots, s_{n-1}, 1 - t)$$

$$= (f \# \bar{f})(s_1, \ldots, s_{n-1}, \frac{1-t}{2}).$$

Thus, by the gluing lemma, 5.73, H is a continuous map, which is constant on the boundary of ∂S_n and hence gives a based homotopy. When $t = 0$, H gives $f \# \bar{f}$ and, when $t = 1$, H sends all points to $(f \# \bar{f})(s_1, \ldots, s_{n-1}, 0)$, i.e., the base point. Hence H is a homotopy from $f \# \bar{f}$ to c.

Picturing H in the usual way, we get

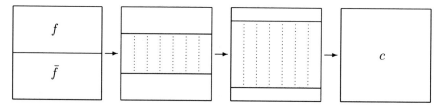

One can prove that $\bar{f} \# f \simeq c$ in a similar way or, by observing that $\bar{\bar{f}} = f$, one can apply what we have proved above to $\bar{f}$, to deduce that $\bar{f} \# \bar{\bar{f}} \simeq c$, i.e., $\bar{f} \# f \simeq c$. $\square$

Proposition 8.4

If f, g, h are pointed maps $S^n \to X$, then $(f \# g) \# h$ is homotopic to $f \# (g \# h)$.
Hence the operation $+$ on $\pi_n(X)$ is associative.

Proof

A homotopy $H : S_n \to X$ is given by

$$H((s_1, \ldots, s_n), t) = \begin{cases} f(s_1, \ldots, s_{n-1}, \frac{4s_n}{2-t}) & \text{if } s_n \leq \frac{2-t}{4}, \\ g(s_1, \ldots, s_{n-1}, 4s_n + t - 2) & \text{if } \frac{2-t}{4} \leq s_n \leq \frac{3-t}{4}, \\ h(s_1, \ldots, s_{n-1}, \frac{4s_n+t-3}{t+1}) & \text{if } \frac{3-t}{4} \leq s_n. \end{cases}$$

As usual, some checking needs to be done: That this gives a consistent definition when $s_n = (3-t)/4$ or $s_n = (2-t)/4$, that $t = 0$ gives the map $(f\#g)\#h$ and that $t = 1$ gives $f\#(g\#h)$. The details are omitted, being very similar to the preceding proofs. This homotopy H can be depicted in the following way.

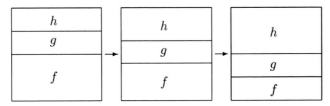

Definition: Let n be any positive integer. The nth **homotopy group** of a pointed space X is the group of pointed homotopy classes of maps $S^n \to X$ with the group operation $[f] + [g] = [f\#g]$.

Example 8.5

If $X \subset \mathbf{R}^n$ is convex, then every pointed map $f : S^n \to (X, x_0)$ is homotopic to the constant map by

$$H(\mathbf{x}, t) = tf(\mathbf{x}) + (1 - t)x_0,$$

where $\mathbf{x} \in S^n$. This homotopy respects the base point: $H((1, 0, \ldots, 0), t) = tf(1, 0, \ldots, 0) + (1 - t)x_0 = tx_0 + (1 - t)x_0 = x_0$ for all t, so f gives the same class in $\pi_n(X)$ as the constant map to x_0. Hence $\pi_n(X) = \{0\}$.

A map which is homotopic to the constant map is said to be **null-homotopic**. This example then shows that any map from S^n to a convex subset of $\mathbf{R}^n$ is null-homotopic.

Example 8.6

If $X = S^1$, then $\pi_1(S^1) = \mathbf{Z}$, since path lifting and homotopy lifting can be carried out in the pointed setting just as well as in the unpointed setting.

Moreover, $\pi_2(S^1) = 0$ which can be seen as follows. Consider a map $S^2 \to S^1$ as a map $f : (S_2, \partial S_2) \to (S^1, (1,0))$ as usual. This can be lifted, by Proposition 6.28, to a map $\tilde{f} : S_2 \to \mathbf{R}$. Since $f(s_1, s_2)$ is the base point of S^1 whenever (s_1, s_2) is on the boundary ∂S_2, so $\tilde{f}(s_1, s_2) \in \mathbf{Z}$ for all such (s_1, s_2). This boundary is connected so, by Lemma 4.18, $\tilde{f}$ is constant on the boundary. (This is where the case $n = 1$ differs: The boundary of $[0,1]$ is not connected.) Since $\mathbf{R}$ is convex, we can define a homotopy $\tilde{F} : S_2 \times [0,1] \to \mathbf{R}$ by

$$\tilde{F}(s_1, s_2, t) = t\tilde{f}(s_1, s_2) + (1 - t)\tilde{f}(0,0).$$

This is continuous and we can compose this with the exponential map e to get a map $F : S_2 \times [0,1] \to S^1$. When $t = 0$, $F(s_1, s_2, 0) = e(\tilde{f}(0,0))$ is constant, whereas when $t = 1$, $F(s_1, s_2, t) = e(\tilde{f}(s_1, s_2)) = f(s_1, s_2)$. Hence F is a homotopy from f to a constant map. Moreover, since $\tilde{f}(s_1, s_2)$ is constant when (s_1, s_2) is in ∂S_2, we have

$$F(s_1, s_2, t) = e(t\tilde{f}(s_1, s_2) + (1 - t)\tilde{f}(0,0)) = e(t\tilde{f}(0,0) + (1 - t)\tilde{f}(0,0))$$
$$= e\tilde{f}(0,0) = f(0,0)$$

for all $(s_1, s_2) \in \partial S_2$. In other words, F corresponds to a pointed homotopy $S^2 \times [0,1] \to S^1$ from the original map $S^2 \to S^1$ to the constant map to the base point. Hence all maps $S^2 \to S^1$ are homotopic.

The same argument can be worked out for all $n > 1$, showing that $\pi_n(S^1) = 0$ unless $n = 1$.

A space such as this, with only one non-trivial homotopy group, is called an **Eilenberg–MacLane space**. If n is a positive integer, and G is a group, then we write $K(G, n)$ for a space with the properties that $\pi_n(K(G, n)) = G$ and $\pi_i(K(G, n)) = 0$ if $i \neq n$. So we could write $K(\mathbf{Z}, 1)$ for S^1. In fact an Eilenberg–MacLane space $K(G, n)$ exists for all combinations of n and G, provided that G is Abelian if $n > 1$ (because of Theorem 8.17 below). This notation is a little misleading, because there can be different spaces with the same homotopy groups, e.g., $\mathbf{R}^2 - \{0\}$ would also be written as $K(\mathbf{Z}, 1)$. However, $\mathbf{R}^2 - \{0\}$ is homotopy equivalent to S^1, and it can be shown that this always happens: Given n and G, any two $K(G, n)$'s are homotopy equivalent. (See Section 4.2 of [5] for a proof of this result.) So, as far as homotopy is concerned, there is a **unique** Eilenberg–MacLane space $K(G, n)$. In the case $n = 1$, an Eilenberg–MacLane space $K(G, 1)$ is called a **classifying space**[1] for the group G and is often denoted by BG. So, for example, S^1 is a classifying space for $\mathbf{Z}$, i.e., $B\mathbf{Z} = S^1$.

---

[1] What does a classifying space classify? It classifies fibre bundles – see [6] for all the details.

The following example misleadingly suggests that the other spheres S^n may also be Eilenberg–MacLane spaces.

Example 8.7

If $n > 1$, then $\pi_i(S^n) = 0$ for $i < n$, and $\pi_n(S^n) = \mathbf{Z}$. We will not be able to prove this rigorously, but the first assertion follows from the fact that a map from an i-dimensional space into a space of higher dimension can be deformed, by a homotopy, into a map which is not surjective. (A proof of this fact can be found in Section 4.1 of [5].) Then, given any map to S^n which is not surjective, we can use stereographic projection away from any point not in its image, and consider it as a map to $\mathbf{R}^n$. All such maps are null-homotopic, and hence the original map to S^n is null-homotopic as well.

The second assertion is proved using the Freudenthal suspension theorem in Chapter 11.

If X is a space like S^n, with $\pi_i(X) = 0$ for $i < n$, then we say that X is $n - 1$-**connected**.

However, if $i > n$, then $\pi_i(S^n)$ can be non-zero, as in the following example.

Example 8.8

The group $\pi_3(S^2)$ is isomorphic to $\mathbf{Z}$. This was discovered by Hopf in the 1930s and represents the beginning of the subject of homotopy theory. Hopf constructed a map $H : S^3 \to S^2$, which can be described as follows. Identifying $\mathbf{R}^4$ with $\mathbf{C}^2$, S^3 corresponds to the set of pairs (z_1, z_2) satisfying $|z_1|^2 + |z_2|^2 = 1$. Division gives an operation $(z_1, z_2) \mapsto z_1/z_2$ which takes values in $\mathbf{C} \cup \{\infty\}$, where $z_1/0$ is understood as ∞. Identifying $\mathbf{C} \cup \{\infty\}$ with the sphere S^2 as in Example 5.7, we end up with a continuous function $H : S^3 \to S^2$.

The preimage of any point in S^2 under H is homeomorphic to a circle S^1, and we will see in Chapter 11 how to use this to show that $\pi_3(S^2) = \pi_3(S^3) = \mathbf{Z}$.

In fact, $\pi_i(S^n)$ is *usually* non-zero if $i > n$, but calculating such homotopy groups of spheres is ridiculously hard, and the results are only known for a small range of i and n. At the time of writing topologists do not hold out much hope of calculating all the homotopy groups of spheres in the foreseeable future.

Given how hard it can be to calculate homotopy groups directly, the following lemma is tremendously useful.

Lemma 8.9

If X, Y are any pointed spaces, then $\pi_n(X \times Y) = \pi_n(X) \times \pi_n(Y)$.

Proof

From Theorem 5.42, we know that a continuous map $f : S^n \to X \times Y$ corresponds to a pair of continuous maps $f_X : S^n \to X$, $f_Y : S^n \to Y$. Suppose that $g : S^n \to X \times Y$ is another map, corresponding to $g_X : S^n \to X$, $g_Y : S^n \to Y$, and $F : S^n \times [0,1] \to X \times Y$ is a homotopy between f and g. Then F corresponds to a pair of maps $F_X : S^n \times [0,1] \to X$ and $F_Y : S^n \times [0,1] \to Y$, and it is easy to see that F_X is a homotopy from f_X to g_X and F_Y is a homotopy from f_Y to g_Y.

Similarly, if we have two homotopic maps $f_X : S^n \to X$ and $g_X : S^n \to X$ and two homotopic maps $f_Y : S^n \to Y$ and $g_Y : S^n \to Y$, then these homotopies combine to give a homotopy between the maps $S^n \to X \times Y$ which correspond to the two pairs (f_X, f_Y) and (g_X, g_Y).

Now, these correspondences work just as well with pointed maps and pointed homotopies: A pointed map $f : S^n \to X \times Y$ corresponds to a pair of pointed maps $f_X : S^n \to X$, $f_Y : S^n \to Y$ and so on. Thus there is a one-to-one correspondence between $\pi_n(X \times Y)$ and $\pi_n(X) \times \pi_n(Y)$.

Finally, this correspondence preserves the addition operations on these groups, which can be seen fairly easily from the way addition on $\pi_n(\ )$ is defined. Hence the group $\pi_n(X \times Y)$ is isomorphic to $\pi_n(X) \times \pi_n(Y)$. □

Example 8.10

The cylinder $C = S^1 \times I$ is the product of S^1 and a contractible space, hence $\pi_n(C) = \pi_n(S^1) \times \pi_n(I) = \pi_n(S^1)$.

Example 8.11

The homotopy groups of the torus $T^2 = S^1 \times S^1$ are $\pi_1(T^2) = \mathbf{Z} \times \mathbf{Z}$, $\pi_n(T^2) = 0$ if $n > 1$. Hence, by Proposition 8.15, S^1 and T^2 are not homotopy equivalent.

8.2 Induced Homomorphisms

The homotopy groups of two homeomorphic spaces are, as you would expect, isomorphic. But the way that homotopy groups are built, directly from sets of continuous maps, leads to a much stronger fact, namely that continuous maps induce homomorphisms of homotopy groups.

Definition: Given a pointed continuous map $f : X \to Y$, there is an **induced function**, $f_* : \pi_n(X) \to \pi_n(Y)$, defined by $f_*[j] = [f \circ j]$ for any pointed map $j : S^n \to X$.

Since $f_*[j]$ is defined in terms of j, we need to verify that f is **well defined**, meaning that if we replace j by another map k homotopic to j (so that $[j] = [k]$), then $f_*[k] = f_*[j]$. This follows directly from Proposition 6.9, since $f \circ k \simeq f \circ j$.

Theorem 8.12

Given a pointed continuous map $f : X \to Y$, the induced function $f_* : \pi_n(X) \to \pi_n(Y)$ is a group homomorphism, with the following properties:

1. If $g : Y \to Z$ is another pointed map, then $(g \circ f)_* = g_* \circ f_*$.

2. If $i : X \to X$ is the identity map, then i_* is the identity homomorphism $\pi_n(X) \to \pi_n(X)$ for each n.

3. If $h : X \to Y$ is (pointed) homotopic to f, then $h_* = f_*$.

4. If $c : X \to Y$ takes every point of X to the base point of Y, then $c_* = 0$, the zero homomorphism.

Proof

To see that f_* is a group homomorphism, let $j_1, j_2 : S^n \to X$ be any pointed maps. Now $f_*([j_1] + [j_2]) = f_*([j_1 \# j_2]) = [f \circ (j_1 \# j_2)]$. If we think of j_1 and j_2 as maps from $(S_n, \partial S_n)$, then

$$f \circ (j_1 \# j_2)(s_1, \ldots, s_n) = \begin{cases} f(j_1(s_1, \ldots, s_{n-1}, 2s_n)) & \text{if } s_n \leq \frac{1}{2}, \\ f(j_2(s_1, \ldots, s_{n-1}, 2s_n - 1)) & \text{if } s_n \geq \frac{1}{2} \end{cases}$$

for $(s_1, \ldots, s_n) \in S_n$, and the expression on the right-hand side here is exactly the same as for $(f \circ j_1) \# (f \circ j_2)$. Hence $[f \circ (j_1 \# j_2)] = [(f \circ j_1) \# (f \circ j_2)] = [f \circ j_1] + [f \circ j_2] = f_*[j_1] + f_*[j_2]$. The properties of this group homomorphism are proved as follows.

1. Since the function f_* is defined by composition with f, it is clear that following this with composition by g will give the same function as composition with $g \circ f$.

2. If $i : X \to X$ is the identity map, and $j : S^n \to X$ is any map, then $i_*[j] = [i \circ j] = [j]$, since $i \circ j = j$. Thus the induced homomorphism i_* is the identity function on $\pi_n(X)$.

3. This follows from Proposition 6.9.

4. Composing any map $j : S^n \to X$ with c will give the constant map to the base-point, which represents the zero element of $\pi_n(Y)$. Hence $c_*([j]) = 0 \in \pi_n(Y)$ for all $[j] \in \pi_n(X)$.   □

Because of Theorem 8.12, f_* is often called the **induced homomorphism** on homotopy groups. One example of the use of induced homomorphisms is in the following result.

Proposition 8.13

There is no continuous map $f : D^2 \to S^1$ such that $f(x, y) = (x, y)$ for all (x, y) in S^1.

Proof

If such a map f existed, then we could combine it with the inclusion $i : S^1 \to D^2$ of S^1 onto the boundary of D^2 in a sequence

$$S^1 \xrightarrow{\;i\;} D^2 \xrightarrow{\;f\;} S^1.$$

Applying π_1 to this sequence, we get a sequence of groups and group homomorphisms

$$\pi_1(S^1) \xrightarrow{\;i_*\;} \pi_1(D^2) \xrightarrow{\;f_*\;} \pi_1(S^1).$$

We know that $\pi_1(S^1) = \mathbf{Z}$ and $\pi_1(D^2) = 0$ since D^2 is convex, hence this sequence is actually

$$\mathbf{Z} \xrightarrow{\;i_*\;} 0 \xrightarrow{\;f_*\;} \mathbf{Z}.$$

Hence, if we take any integer in $\mathbf{Z}$ and apply i_* and then apply f_*, we will get the answer $f_*(0)$, since $i_*(n) = 0$ for all n.

However, the condition that $f(x, y) = (x, y)$ for $(x, y) \in S^1$ says that $f \circ i$ is the identity on S^1. Thus $f_* \circ i_* = (f \circ i)_*$ is the identity group homomorphism on $\pi_1(S^1) = \mathbf{Z}$. Yet we have just seen that, because this homomorphism factors through the 0 group, it must be the zero homomorphism, not the identity. This contradiction shows that such an f cannot exist.   □

Note from this proof that the inclusion map $i : S^1 \to D^2$ induces the zero homomorphism on π_1. Thus a continuous map which is an injection need not induce an injective homomorphism on homotopy groups. Similarly a surjective map need not induce a surjective homomorphism.

Proposition 8.13 is a typical example of how induced homomorphisms are used. Note that we did not use any information other than that given by Theorem 8.12. Theoretically, one could take a description of the continuous map and derive an explicit description of the induced homomorphism, but this is rarely worth the effort. It is better to let the algebra do the work, using the information given by Theorem 8.12 and the knowledge we have of group homomorphisms $\pi_n(X) \to \pi_n(Y)$, just as in the proof of the following proposition.

Proposition 8.14

If $f : S^1 \to S^1$ is any map such that $f \circ f$ is the constant map to the base point, then $f_* : \pi_1(S^1) \to \pi_1(S^1)$ is the zero homomorphism.

Proof

The composite $f \circ f$ induces $(f \circ f)_* = f_* \circ f_*$. Since $f \circ f$ is constant, this induced map must be the zero homomorphism, by part 4 of Theorem 8.12. Now $\pi_1(S^1) = \mathbf{Z}$ and the only group homomorphism $\alpha : \mathbf{Z} \to \mathbf{Z}$ such that $\alpha \circ \alpha = 0$ is the zero homomorphism. Hence $f_* = 0$. ☐

Theorem 8.12 also implies that homotopy groups cannot distinguish between homotopy equivalent spaces:

Proposition 8.15

If S and T are (pointed) homotopy equivalent, then $\pi_n(S)$ and $\pi_n(T)$ are isomorphic for all n.

Proof

Let $f : S \to T$ and $g : T \to S$ be continuous maps whose composites are homotopic to the identity. The induced homomorphisms f_* and g_* are then inverse to each other. For example, the composite $f \circ g$ is homotopic to the identity on T and so induces the identity, i.e., $(f \circ g)_* = id$. But this is $f_* \circ g_*$, so $f_* \circ g_* = id$. Similarly, $g_* \circ f_* = id$, i.e., both f_* and g_* are invertible, hence they are isomorphisms. ☐

Example 8.16

The space $\mathbf{R}^2 - \{0\}$ is homotopy equivalent to S^1, hence $\pi_i(\mathbf{R}^2 - \{0\}) = 0$ unless $i = 1$, and $\pi_1(\mathbf{R}^2 - \{0\}) = \mathbf{Z}$.

The proposition also confirms that homeomorphic spaces have the same homotopy groups, since they are homotopy equivalent.

8.3 The Fundamental Group

For a long time the group $\pi_1(X)$ was felt to be much more important than the other homotopy groups, and was named the **fundamental group**. Its relative importance was due to the following result.

Theorem 8.17

If $n > 1$, then $\pi_n(X)$ is an Abelian group.

Proof

The homotopy is a little more complicated than preceding examples and is best considered in stages. Rather than give formulae, we will just draw the pictures, suppressing all but s_1, s_n as usual. The symbol c denotes the constant map to the base point of X.

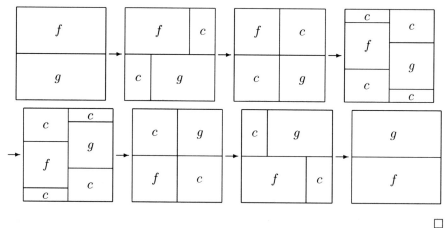

The proof clearly uses both the s_1-coordinate and the s_n-coordinate, and the result is not true for $n = 1$; in general, the fundamental group is not Abelian. For some spaces, $\pi_1(X)$ *is* Abelian, but this is a reflection of the topology of the space X.

Theorem 8.17 led early topologists to feel that the "higher" homotopy groups could not contain much information, unlike the fundamental group.

However, they soon discovered the fallacy of this when they found out how complicated the homotopy groups of spheres could be.

The simplest space with non-Abelian fundamental group is the figure of eight:

Example 8.18

Let $X \subset \mathbf{R}^2$ be the space of points (x, y) such that $(|x| - 1)^2 + y^2 = 1$, i.e., X is a figure of eight:

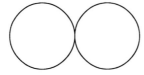

Define $f : S^1 \to X$ by $f(x, y) = (1-x, y)$ and $g : S^1 \to X$ by $g(x, y) = (x-1, y)$, so that f loops round the right-hand circle, and g around the left-hand circle. These two maps can be combined in two different ways: We can take $f \# g$ or we can take $g \# f$. Although it is not straightforward, it can be shown (e.g., in Section 1.2 of [5], or Chapter IV of [7]) that these two sums are *not* homotopic, hence $\pi_1(X)$ is not Abelian.

8.4 Path Connectivity and π_0

The addition operation we defined on $[S^n, X]$ for $n > 0$ does not have an analogue if $n = 0$; the set $\pi_0(X) = [S^0, X]$ does not generally have a group structure. Nevertheless it contains some useful information about X. If we have a pointed map $f : S^0 \to X$, then f is determined by $f(-1)$, since $f(1)$ must be the base point of X. Two maps $f, g : S^0 \to X$ are pointed homotopic if there is a path in X from $f(-1)$ to $g(-1)$. This leads us to a slightly more restrictive notion of connectivity:

> Definition: A space X is **path connected** if, given any two points $x, y \in X$, there is a continuous map $p : [0, 1] \to X$ such that $p(0) = x$ and $p(1) = y$.

Lemma 8.19

A space X is path connected if, and only if, $\pi_0(X)$ has only one element.

Proof

By the comments made above, two maps $f, g : S^0 \to X$ are (pointed) homotopic if there is a path from $f(-1)$ to $g(-1)$. If X is path connected, then this will always be the case no matter what f and g are. In other words, all maps $S^0 \to X$ are homotopic.

Conversely, if x, y are any two points in X, then we can define continuous maps $f, g : S^0 \to X$ by $f(-1) = x$, $g(-1) = y$. Since $\pi_0(X)$ has only one element, these maps are homotopic, hence there is a path from x to y. So X is path connected. $\qquad\square$

If a space is not path connected, then we can form an equivalence relation on the points in X, where $x \sim y$ if there is a path from x to y. The set of equivalence classes is exactly the set of homotopy classes $[S^0, X]$. For convenience this is often dealt with alongside the homotopy groups, and even sometimes called the 0th homotopy group, even though it is not a group!

Note that path connectivity is a stronger condition than the notion of connectivity introduced in Chapter 4.

Proposition 8.20

If X is path connected, then X is connected.

Proof

Suppose that X is disconnected, say $X = U \cup V$ where $U \cap V = \emptyset$ and U and V are non-empty open sets. Let $x \in U$ and $y \in V$. Since X is path connected, there is a continuous map $p : [0, 1] \to X$ such that $p(0) = x$, $p(1) = y$. As U and V are open, their preimages $p^{-1}(U)$ and $p^{-1}(V)$ are open subsets of $[0, 1]$. Neither of these preimages is empty, since one contains 0 and the other contains 1. And since $U \cup V = X$, so $p^{-1}(U) \cup p^{-1}(V) = [0, 1]$. This says that $[0, 1]$ is disconnected, which we know to be false, from Example 4.5. This contradiction shows that X must be connected. $\qquad\square$

Conversely, a connected space *need not* be path connected. The classical example of this is the following.

Example 8.21

Let $X \subset \mathbf{R}^2$ be the set

$$X = \{(x, y) : x = 0, -1 \le y \le 1\} \cup \{(x, y) : 0 < x \le 1 \text{ and } y = \sin(1/x)\}.$$

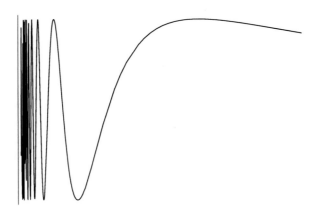

Suppose that $p : [0,1] \to X$ is a path such that $p(0) = (0,0)$ and $p(1) = (1, \sin(1))$. Then

$$S = \{(x,y) : 0 < x \le 1 \text{ and } y = \sin(1/x)\}$$

is an open subset of X, so its preimage under p is an open set of $[0,1]$. Thus, this can be written as a union of basic open sets, and this will include an interval $(b,1]$ for some $b \in [0,1]$. Assuming b is chosen as small as possible, such that $(b,1] \subset p^{-1}(S)$, we see that $p(b) \notin S$. (Otherwise $b \in p^{-1}(S)$ and there will be an open interval around b contained in $p^{-1}(S)$ that we can use to enlarge $(b,1]$.)

Now suppose that $p(b) = (0,y)$ where $y \le 0$. Let $\delta > 0$ be any number such that $y + \delta < 1$ and consider the open ball of radius δ about $p(b)$. Since this is open, its preimage is also open, and so contains an interval $[b, b + \epsilon)$ for some $\epsilon > 0$. Since $(b,1] \subset p^{-1}(S)$, we know that $p(b + \epsilon) \in S$. So $p(b + \epsilon) = (x', y')$ where $x' > 0$. For any $x' > 0$, there is some number of the form $2/\pi(4n+1)$ less than x', with n an integer, so that $\sin(1/(2/\pi(4n+1))) = \sin(\pi(4n+1)/2) = 1$. By the intermediate value theorem, there is some $t \in (b, b+\epsilon)$ such that $p(t) = (2/\pi(4n+1), y'')$ for some y''. Since $p(t) \in S$, $y'' = \sin(1/(2/\pi(4n+1))) = 1$ and, consequently $p(t) \notin B_\delta(p(b))$, even though $t \in (b, b+\epsilon) \subset p^{-1}(B_\delta(p(b)))$. Thus we have a contradiction.

Similarly, if $p(b) = (0,y)$ where $y > 0$, we can arrive at a contradiction. Thus, either way, there can be no such path p. So X is not path connected.

However, X *is* connected. To see this, suppose that $X = U \cup V$ for two non-empty, open, disjoint subsets U and V. First consider the subset $I = \{(x,y) : x = 0, -1 \le y \le 1\}$ of X as a subspace. This is homeomorphic to $[0,1]$, and so is connected. If we take the intersections $I \cap U$ and $I \cap V$ then, in the subspace topology, we get two disjoint open sets which cover I. Since I is connected, one of these must be empty. Let V be such that $I \cap V$ is empty, so $I \subset U$. On the other hand, let $J = X - I = \{(x,y) : 0 < x \le 1, y = \sin(1/x)\}$. This is homeomorphic to $(0, 1]$ and so is connected as well. By a similar argument, we must have $J \cap U = \emptyset$ and $J \subset V$, i.e., $U = I$, $V = J$. However, I is not open in X; any open set of $\mathbf{R}^2$ containing I will also contain some points of J. Hence there can be no sets U and V satisfying our assumptions, i.e., X is connected.

Path connectivity behaves in many formal ways like connectivity. For example, the image of a path connected space under a continuous map is path connected. Moreover, a space can be split into **path components** since this is based on an equivalence relation. However, a space is not generally the disjoint union of its path components, as Example 8.21 shows. A more extreme example of such behaviour is the space $\mathbf{Q}$ whose path components consist of single points, yet, as Section 5.2 shows, $\mathbf{Q}$ is not homeomorphic to the disjoint union of its elements.

Since the image of a path connected space under a continuous map is path connected, and S^n is path connected if $n > 0$ (Exercise 8.3), homotopy groups are of limited use in studying non-path connected spaces.

Proposition 8.22

If $n > 0$ and X is any pointed topological space, then $\pi_n(X) = \pi_n(X_0)$, where X_0 is the path component of X which contains the base point.

Corollary 8.23

The homotopy groups of $\mathbf{Q}$ are as follows: $\pi_0(\mathbf{Q}) = \mathbf{Q}$, $\pi_i(\mathbf{Q}) = 0$ for $i > 0$.

Similarly, $\pi_0(\mathbf{Z}) = \mathbf{Z}$ and $\pi_i(\mathbf{Z}) = 0$ for $i > 0$. Since $\mathbf{Z}$ and $\mathbf{Q}$ are isomorphic as sets, we could equally well say $\pi_0(\mathbf{Z}) = \mathbf{Q}$. So homotopy groups are unable to distinguish the intricate topological space $\mathbf{Q}$ from the relatively simple space $\mathbf{Z}$.

8.5 The Van Kampen Theorem

It is very hard, in general, to calculate the homotopy groups of a space. One useful tool is the Van Kampen theorem which describes the fundamental group of a space in terms of the fundamental group of its subspaces, subject to certain conditions.

Theorem 8.24 (Van Kampen Theorem)

Suppose that $X = U \cup V$ where U and V are open subsets of X such that $U \cap V$ is path connected and contains the base point of X. Then every element $\alpha \in \pi_1(X)$ can be written as a sum

$$\alpha = \beta_1 + \beta_2 + \cdots + \beta_n,$$

where, for each i, either $\beta_i \in j_*(\pi_1(U))$ or $\beta_i \in k_*(\pi_1(V))$, where $j_* : \pi_1(U) \to \pi_1(X)$ and $k_* : \pi_1(V) \to \pi_1(X)$ are the homomorphisms induced by the inclusions $j : U \subset X$, $k : V \subset X$.

Proof

Suppose that $\alpha \in \pi_1(X)$ is represented by $f : S^1 \to X$. As usual, we think of this as a map $f : [0, 1] \to X$ with $f(0) = f(1)$ equal to the base point of X.

We will first look at a very simple case, before seeing how to generalize this. We suppose that $f(t) \in U$ if $t \leq 1/2$ and $f(t) \in V$ if $t \geq 1/2$. In other words, $f[0, 1/2] \subset U$ and $f[1/2, 1] \subset V$. This shows that $f(1/2) \in U \cap V$. As $U \cap V$ is assumed to be path connected, there is some path $g : [0, 1] \to U \cap V$ such that $g(0) = f(1/2)$ and $g(1)$ is the base point in X (which is in $U \cap V$, by assumption). Now we define two paths, $f_1, f_2 : [0, 1] \to X$, by:

$$f_1(s) = \begin{cases} f(s) & \text{if } s \leq \frac{1}{2}, \\ g(2s - 1) & \text{if } s \geq \frac{1}{2}, \end{cases} \quad \text{and} \quad f_2(s) = \begin{cases} g(1 - 2s) & \text{if } s \leq \frac{1}{2}, \\ f(s) & \text{if } s \geq \frac{1}{2}. \end{cases}$$

These are continuous, by the gluing lemma 5.73, since $f(1/2) = g(0)$.

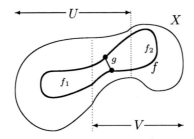

As the picture makes clear, we have arranged f_1 and f_2 so that the image of f_1 is contained in U and the image of f_2 is contained in V. In other words, $f_1 = j \circ g_1$ and $f_2 = k \circ g_2$ for some paths $g_1 : [0, 1] \to U$ and $g_2 : [0, 1] \to V$. Hence $[f_1] \in j_*(\pi_1(U))$ and $[f_2] \in k_*(\pi_1(V))$.

We can define a homotopy from $f_1 \# f_2$ to f as follows:

$$F(s, t) = \begin{cases} f_1 \# f_2 \left(\frac{s}{1+t} \right) & \text{if } s \leq \frac{1}{2}, \\ f_1 \# f_2 \left(\frac{s+t}{1+t} \right) & \text{if } s \geq \frac{1}{2}. \end{cases}$$

This is continuous since, when $s = 1/2$, unravelling the definitions we find that both definitions give $F(1/2, t) = g(\frac{1-t}{1+t})$. If $t = 0$, then $F(s, 0) = (f_1 \# f_2)(s)$ and, if $t = 1$,

$$F(s, 1) = \begin{cases} f_1 \# f_2 \left(\frac{s}{2}\right) = f(s) & \text{if } s \leq \frac{1}{2}, \\ f_1 \# f_2 \left(\frac{s+1}{2}\right) = f(s) & \text{if } s \geq \frac{1}{2}. \end{cases}$$

Thus F is a homotopy from $f_1 \# f_2$ to f. Hence $\alpha = \beta + \gamma$, where $\beta \in j_*(\pi_1(U))$ is the class of f_1, and $\gamma \in k_*(\pi_1(V))$ is the class of f_2.

In terms of the previous picture, we have broken up the loop f as

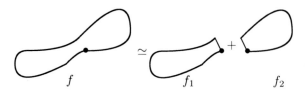

$$f \qquad\qquad\qquad f_1 \qquad\qquad\qquad f_2$$

Clearly, by adapting this argument, we could replace $1/2$ by any number $x \in [0, 1]$ such that $f(t) \in U$ for $t \leq x$ and $f(t) \in V$ if $t \geq x$. We can also adapt this argument to handle a situation where there are, say, two numbers x_1, x_2, such that $f(t) \in U$ if $t \leq x_1$, $f(t) \in V$ if $x_1 \leq t \leq x_2$ and $f(t) \in U$ if $t \geq x_2$. Similarly, we can handle any finite number of such x_i's, where f alternates between U and V, i.e., if $f[x_i, x_{i+1}] \subset U$, then $f[x_{i-1}, x_i] \subset V$ and $f[x_{i+1}, x_{i+2}] \subset V$.

By the domain splitting result, Proposition 6.29, we can always find some n such that each subinterval $[i/n, (i+1)/n]$ is mapped into either U or V, for each $i = 0, \ldots, n-1$. It need not be the case that adjacent intervals are mapped to different subsets U and V but, by merging adjacent intervals as necessary, we can obtain a list of intervals with this property.

Hence any map $S^1 \to X$ is homotopic to a $\#$ sum of maps $S^1 \to U \to X$ and maps $S^1 \to V \to X$, which proves the theorem. $\qquad\qquad\square$

Example 8.25

We claimed earlier that S^n is $n-1$-connected, i.e., $\pi_i(S^n) = 0$ if $i < n$. The Van Kampen theorem allows us to see part of this (that $\pi_1(S^n) = 0$) quite easily.

Write S^n as the union $S^n = U \cup V$ where $U = S^n - \{(0, \ldots, 0, 1)\}$ and $V = S^n - \{(0, \ldots, 0, -1)\}$, so both U and V omit one point. By stereographic projection, U and V are homeomorphic with $\mathbf{R}^n$ and, consequently, contractible. So $\pi_1(U) = \pi_1(V) = 0$. The intersection $U \cap V$ is homeomorphic to an open cylinder $S^{n-1} \times (-1, 1)$ and homotopy equivalent to S^{n-1}. This is certainly path connected, so Van Kampen's theorem applies, and we see that $\pi_1(S^n) = 0$.

If U and V have trivial fundamental group, then the Van Kampen theorem shows that $\pi_1(X)$ is also trivial. However, if U and V have non-trivial fundamental group, then this theorem does not completely determine $\pi_1(X)$, as the following examples show.

Example 8.26

Let $X = S^1$ and let $U = V = X$. Then $\pi_1(U) = \pi_1(V) = \mathbf{Z}$ and $\pi_1(X) = \mathbf{Z}$.

On the other hand, let X be the disc, $X = D^2$, and let $U = D^2 - \{(-1,0)\}$ and $V = D^2 - \{(1,0)\}$. Then U and V are homotopy equivalent to S^1, so $\pi_1(U) = \pi_1(V) = \mathbf{Z}$ as before. But X is contractible, so $\pi_1(X) = 0$.

Thus, we cannot hope to determine $\pi_1(X)$ precisely from knowledge of $\pi_1(U)$ and $\pi_1(V)$ alone.

The problem here is that we must take account of the intersection $U \cap V$. There is a stronger version of the Van Kampen theorem which explains how this intersection affects the fundamental group of X. Unfortunately, even the statement of this result is complicated, so we refer the reader to Section 1.2 of [5], or Section 14c of [4], for the details.

Despite how hard they are to calculate, homotopy groups are very useful because they capture so much topological information. One example of this is the following theorem, a proof of which can be found in Section 4.1 of [5].

Theorem 8.27 (Whitehead Theorem)

If X and Y are connected simplicial complexes and there is a map $f : X \to Y$ such that $f_* : \pi_i(X) \to \pi_i(Y)$ is an isomorphism for all i, then f is a homotopy equivalence between X and Y, i.e., there is a map $g : Y \to X$ such that $f \circ g$ and $g \circ f$ are homotopic to the respective identity maps.

Note that this does *not* say that if X and Y have the same homotopy groups then they are homotopy equivalent. This is not true, and the Whitehead theorem needs there to be a map inducing isomorphisms of homotopy groups. But it does say that homotopy groups, along with induced homomorphisms, reflect topological situations well.

EXERCISES

8.1. Complete the proof that $\pi_n(S^1) = 0$ if $n > 1$.

8.2. Prove that there can be no continuous surjection from a path connected space to a space which is not path connected.

8.3. Prove that S^n is path connected for $n \geq 1$.

8.4. Let $f : S^1 \to S^1$ be a map of degree n. Describe the induced homomorphism $f_* : \pi_1(S^1) \to \pi_1(S^1)$.

8.5. Calculate the homotopy groups of the complement $\mathbf{R}^2 - S^1$.

9
Simplicial Homology

In Chapter 8, we have seen how to take a topological space and assign to it an algebraic object carrying some information about the topology of the space. However, homotopy groups are very hard to calculate. In this chapter we introduce "homology" groups, which can be thought of as a rough approximation to the homotopy groups of a space. In defining the homotopy groups of a space X, we considered all maps $f : S^n \to X$ and ignored those which could be deformed to a constant map. Such a deformation means extending f to a map $D^n \to X$, i.e., filling in the image of S^n under f. So the nth homotopy group could be roughly described as counting those images of S^n in X which *cannot* be filled in. We could think of such images as "n-dimensional holes," so that the annulus has a one-dimensional hole, i.e., a hole which can be bounded by a one-dimensional rope. A 2-sphere would then have a two-dimensional hole, as the hole inside it cannot be bounded by a one-dimensional rope, but has a two-dimensional boundary.

However, this way of understanding $\pi_n(X)$ breaks down when we realize that $\pi_m(S^n)$ is usually not zero if $m > n$, suggesting that the n-sphere usually has lots of m-dimensional holes!

Homology groups offer a different approach to hole counting, and one that behaves slightly more intuitively. For example, with this approach, the n-sphere has one n-dimensional hole and no m-dimensional holes for $m \neq n$ (except $m = 0$ which is an exceptional case).

As with the Euler characteristic, we will begin by defining homology for simplicial complexes before giving a more general definition in Chapter 10. In the simplicial context, a "hole" is some combination of simplices which could

M.D. Crossley, *Essential Topology*, Springer Undergraduate
Mathematics Series, DOI 10.1007/978-1-84628-194-5_9,
© Springer-Verlag London Limited 2010

possibly be the boundary of a simplex, or of a combination of simplices, but which is not. For example, the three edges that form the simplicial circle of Example 7.1 look just like the three edges that form the boundary of a 2-simplex. Hence these edges *could* be a boundary, but in the simplicial circle they are not a boundary, since there is no 2-simplex with this collection of edges as its boundary.

9.1 Simplicial Homology Modulo 2

Describing the boundary of a simplex, or combination of simplices, is essential to homology, so we will begin by recalling what we mean by taking the boundary of a simplex, and we will turn this process into an algebraic function.

Recall that a k-simplex $[v_0, \ldots, v_k]$ has $k+1$ faces, each of which is a $(k-1)$-simplex, given by omitting one of the vertices. There is a very useful convention for describing a list in which one item has been omitted – you put a hat over the omitted item. For example, $\hat{a}, b, c$ would denote the list b, c, the letter a having been omitted.

With this convention, if $[v_0, \ldots, v_k]$ is a k-simplex, then its faces can be written as $[v_0, \ldots, \hat{v_i}, \ldots, v_k]$, where $0 \leq i \leq k$. For example, the faces of a 2-simplex $[a, b, c]$ are

$$[\hat{a}, b, c] = [b, c], \quad [a, \hat{b}, c] = [a, c], \quad [a, b, \hat{c}] = [a, b].$$

The boundary of the k-simplex is the union of these faces, so the boundary of $[a, b, c]$ is the union of $[b, c]$, $[a, c]$ and $[a, b]$, as depicted below.

If, for a given simplicial complex K, we let $S_n(K)$ be the set of all n-simplices in K, then the boundary of each element in $S_n(K)$ is a list of elements of $S_{n-1}(K)$. We would like to construct a "boundary operator" or "boundary function" which takes a simplex and gives its boundary. However, since the boundary is not a single simplex, but a list of simplices, we cannot construct a boundary function $S_n(K) \rightarrow S_{n-1}(K)$. There are various ways to get around this, and the quickest way (though not the best, as we will see later) is as follows. Let $C_n(K)$ be the collection of all subsets of $S_n(K)$, so an element of $C_n(K)$ is a subset of $S_n(K)$. Hence, the boundary of an n-simplex is an element of $C_{n-1}(K)$.

Another way of describing $C_n(K)$ is to say that it is the $\mathbf{Z}/2$-vector space spanned by $S_n(K)$, i.e., the set of all linear combinations

$$\lambda_1\sigma_1 + \cdots + \lambda_k\sigma_k$$

where, for $1 \leq i \leq k$, λ_i is an element of $\mathbf{Z}/2$, i.e., either 0 or 1, and σ_i is an n-simplex in K. Such a linear combination is called an n-**chain** and, by listing those simplices σ_i which have coefficient 1 in this expression, we get a subset of $S_n(K)$. Hence both descriptions agree, but the second definition has the advantage of bringing us into the realm of linear algebra: We can add two n-chains together to get a third n-chain. Note, however, that we may add two n-chains together and get 0 as the answer. For example, $\sigma + \sigma = 2\sigma = 0$ since the coefficients are added in $\mathbf{Z}/2$.

As the boundary of an n-simplex is an element of $C_{n-1}(K)$, we get, for each $n > 0$, a function $d_n : S_n(K) \to C_{n-1}(K)$. We can extend this to a linear transformation $\delta_n : C_n(K) \to C_{n-1}(K)$ by defining

$$\delta_n(\lambda_1\sigma_1 + \cdots + \lambda_k\sigma_k) = \lambda_1 d_n(\sigma_1) + \cdots + \lambda_k d_n(\sigma_k),$$

whenever $\sigma_1, \ldots, \sigma_k$ are n-simplices, and $\lambda_1, \ldots, \lambda_k$ are coefficients in $\mathbf{Z}/2$. This will involve adding up coefficients, which we do modulo 2 of course. For example, if

$$d_n(\sigma_1) = s_1 + s_2 \quad\text{and}\quad d_n(\sigma_2) = s_2 + s_3,$$

where $s_1, s_2, s_3 \in C_{n-1}(K)$, then

$$\delta_n(\sigma_1 + \sigma_2) = (s_1 + s_2) + (s_2 + s_3) = s_1 + s_3.$$

So, for each $n \geq 0$, we have a $\mathbf{Z}/2$-vector space $C_n(K)$ and, for each $n \geq 1$, a linear transformation

$$\delta_n : C_n(K) \longrightarrow C_{n-1}(K)$$

called the **boundary operator.**

For example, applying this to a 1-simplex $[a, b]$ gives

$$\delta_1[a, b] = a + b,$$

and applying it to a 2-simplex $[a, b, c]$, we get

$$\delta_2[a, b, c] = [b, c] + [a, c] + [a, b].$$

In general, the effect of applying δ_n to an arbitrary n-simplex is given by the formula

$$\delta_n[v_0, \ldots, v_n] = \sum_{i=0}^{n}[v_0, \ldots, \widehat{v_i}, \ldots, v_n].$$

The aim of homology is to study those combinations of simplices which *could* be boundaries of simplices (or of a combination of simplices) but which are not. For example, the three innermost edges of the simplicial torus

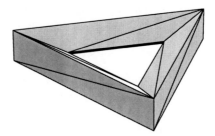

look just like the boundary of a 2-simplex. But there is no 2-simplex in this simplicial torus whose boundary is that combination of edges. In that sense, these three edges *could* be a boundary, but are not.

The problem is to determine which combinations of simplices could be boundaries. Obviously, the combination must look something like the boundary of a simplex, so we will resolve this problem by studying boundaries of simplices.

As we have seen, the boundary of an n-simplex is a union of $(n-1)$-simplices, and these $(n - 1)$-simplices cannot be arbitrary: They are related to each other by the condition that each face of one of these $(n - 1)$-simplices is a face of exactly one other $(n - 1)$-simplex. For example, the boundary of the 2-simplex $[a, b, c]$ consists of the simplices $[b, c]$, $[a, c]$, $[a, b]$. The faces of these boundary simplices are: b, c, a, c and a, b. Each element in this list occurs twice as each face occurs in two boundary simplices. Putting that another way, $\delta_2[a, b, c] = [b, c] + [a, c] + [a, b]$, and

$$\delta_1(\delta_2[a, b, c]) = \delta_1([b, c] + [a, c] + [a, b]) = (b+c) + (a+c) + (a+b) = 2a + 2b + 2c.$$

But, since we are working modulo 2, this is 0. In other words $\delta_1 \circ \delta_2$ is zero on each 2-simplex. This generalizes as follows.

Lemma 9.1

For every $n \geq 1$, the composite

$$\delta_n \circ \delta_{n+1} : C_{n+1}(K) \longrightarrow C_{n-1}(K)$$

is the zero linear transformation.

Proof

Certainly this composite will be a linear transformation, so it is enough to verify it for every element of a basis for $C_{n+1}(K)$. By definition, one basis is $S_{n+1}(K)$, so we will show that $\delta_n(\delta_{n+1}(\sigma)) = 0$ for every $(n + 1)$-simplex σ.

Let $\sigma = [v_0, \ldots, v_{n+1}]$, so

$$\delta_{n+1}(\sigma) = \sum_{i=0}^{n+1} [v_0, \ldots, \widehat{v_i}, \ldots, v_{n+1}].$$

Then

$$\delta_n \delta_{n+1}(\sigma) = \sum_{\substack{j=0 \\ j \neq i}}^{n+1} \sum_{i=0}^{n+1} [v_0, \ldots, \widehat{v_i}, \ldots, \widehat{v_j}, \ldots, v_{n+1}].$$

But every summand here occurs twice: the simplex $[v_0, \ldots, \widehat{v_a}, \ldots, \widehat{v_b}, \ldots, v_{n+1}]$ occurs when $i = a$ and $j = b$, and also when $i = b$ and $j = a$. Since we are working modulo 2, this means that all the summands cancel, leaving

$$\delta_n \delta_{n+1}(\sigma) = 0. \qquad \square$$

Lemma 9.1 shows that we have a sequence of $\mathbf{Z}/2$-vector spaces and linear transformations

$$\cdots \longrightarrow C_n(K) \xrightarrow{\delta_n} C_{n-1}(K) \xrightarrow{\delta_{n-1}} C_{n-2}(K) \longrightarrow \cdots \longrightarrow C_1(K) \xrightarrow{\delta_1} C_0(K),$$

where the composite of any two transformations is 0. Such a sequence is called a **chain complex** and is denoted by $(C_*(K), \delta_*)$, or $C_*(K)$, or even C_* according to context.

So the boundary of an $n + 1$-simplex is in the kernel of δ_n. Traditionally, this is expressed by saying that "a boundary has no boundary." This suggests a way of detecting combinations of simplices which "could" be boundaries: They are those combinations which have no boundary themselves. In other words, they are elements of Ker δ_n, and for $n > 0$, we write $Z_n(K)$ for Ker δ_n, abbreviating this to Z_n if the context makes it clear which simplicial complex we are considering. Elements of Z_n are called **cycles**, the Z coming from the German word for "cycle." For convenience, we define $Z_0(K)$ to be $C_0(K)$.

However, we wish to discard those combinations of simplices which actually are boundaries. These are easily recognised: they are the elements of Im δ_{n+1}. We write $B_n(K)$ (or just B_n) for Im δ_{n+1} and refer to elements of B_n as **boundaries**.

Lemma 9.1 says that $B_n \subset Z_n$ for all $n \geq 0$. So one way to "discard" the cycles which are actually boundaries is to take the **quotient group** Z_n/B_n, whose elements are equivalence classes of cycles under the relation $z_1 \sim z_2$ if $z_1 - z_2$ is in B_n.

Definition: The nth **homology group** of a simplicial complex K is the quotient
$$H_n(K) = \frac{Z_n(K)}{B_n(K)} = \begin{cases} \text{Ker } \delta_n/\text{Im } \delta_{n+1} & \text{if } n > 0, \\ C_0/\text{Im } \delta_1 & \text{if } n = 0. \end{cases}$$
The **homology** of K is the collection
$$H_*(K) = \{H_0(K), H_1(K), H_2(K), \cdots \}.$$

The group $H_n(K)$ is sometimes referred to as the **degree** n part, or **dimension** n part, of the homology of K. Note that since $Z_n(K)$ and $B_n(K)$ are $\mathbf{Z}/2$-vector spaces, with $B_n(K)$ a vector subspace of $Z_n(K)$, $H_n(K)$ is, in fact, a $\mathbf{Z}/2$-vector space, not just a group. However, the term "homology group" is traditional as there is a more common construction, integral homology, which we will meet shortly, which yields an Abelian group, not a vector space.

Since the elements of $H_n(K)$ are equivalence classes of cycles modulo boundaries, we say that two chains z_1 and z_2 are **homologous** if their difference $z_1 - z_2$ is a boundary, i.e., $z_1 - z_2 \in B_n$.

Example 9.2

Let K be the simplicial circle of Example 7.1. This has three 0-simplices, which we label a, b, c, and three 1-simplices, $[a, b]$, $[b, c]$, $[a, c]$. So C_0 and C_1 both have dimension 3, while $C_i = 0$ if $i > 1$. Thus there is only one map in the chain complex which can possibly be non-zero, namely δ_1. So the only interesting part of the chain complex is
$$C_1 \xrightarrow{\delta_1} C_0,$$
where $\delta_1[a, b] = a + b$, $\delta_1[b, c] = b + c$, $\delta_1[a, c] = a + c$. Let $\sigma = \lambda_1[a, b] + \lambda_2[b, c] + \lambda_3[a, c]$ be an arbitrary element of C_1. Then
$$\delta_1(\sigma) = \lambda_1(a + b) + \lambda_2(b + c) + \lambda_3(a + c) = (\lambda_1 + \lambda_3)a + (\lambda_1 + \lambda_2)b + (\lambda_2 + \lambda_3)c.$$

If $\delta_1(\sigma) = 0$, then $\lambda_1 + \lambda_3 = 0$, i.e., $\lambda_3 = \lambda_1$ (since we are working over $\mathbf{Z}/2$, so $-1 = +1$), and $\lambda_1 + \lambda_2 = 0$, i.e., $\lambda_1 = \lambda_2$. In other words, if $\sigma \in \text{Ker } \delta_1$, then
$$\sigma = \lambda([a, b] + [b, c] + [a, c])$$
for some $\lambda \in \mathbf{Z}/2$. Hence $\dim Z_1 = \dim \text{Ker } \delta_1 = 1$. The map δ_2 is 0, so its image B_1 is 0, so $H_1(K) = Z_1/B_1 = Z_1 = \mathbf{Z}/2$.

By the rank-and-nullity theorem of linear algebra, $\dim \text{Im } \delta_1 = \dim C_1 - \dim \text{Ker } \delta_1 = 3 - 1 = 2$. Hence $\dim H_0(K) = \dim(C_0/B_0) = \dim(C_0) - \dim(B_0) = 3 - 2 = 1$.

As $C_i = 0$ for $i > 1$, so $H_i(K) = 0$ for $i > 1$, and we have a complete calculation of the homology groups: $H_0(K) = \mathbf{Z}/2$, $H_1(K) = \mathbf{Z}/2$ and $H_i(K) = 0$ for $i > 1$.

Example 9.3

The simplicial square of Example 7.2 has four 0-simplices, five 1-simplices and two 2-simplices, so the non-zero part of the chain complex is longer:

$$C_2 \xrightarrow{\delta_2} C_1 \xrightarrow{\delta_1} C_0.$$

As in Example 9.2, the image of δ_1 consists of all sums $v_i + v_j$ of two vertices, the sum $v_1 + v_2 + v_3 + v_4$ of all four vertices, and 0. So $B_0 = \operatorname{Im} \delta_1$ contains 8 elements, and $\dim \operatorname{Im} \delta_1$ must be 3. Hence $\dim H_0 = \dim(C_0) - \dim(B_0) = 4 - 3 = 1$.

As $\dim \operatorname{Im} \delta_1 = 3$, the rank-and-nullity theorem shows that $\dim \operatorname{Ker} \delta_1 = 2$. Since the two 2-simplices have different boundaries, so $\dim \operatorname{Im} \delta_2 = 2$, and hence $H_1 = 0$.

Finally, since $\dim \operatorname{Ker} \delta_2 = 0$, we see that $H_2 = 0$ as well. In other words, the square has non-zero homology only in dimension 0. Hence $H_0 = \mathbf{Z}/2$, $H_i = 0$ for $i > 0$.

In this situation, where the only non-zero homology group is in dimension 0, the simplicial complex is said to be **acyclic**.

Example 9.4

For the simplicial torus of Example 7.8, the non-zero part of the chain complex is

$$C_2 \xrightarrow{\delta_2} C_1 \xrightarrow{\delta_1} C_0,$$

where $\dim C_2 = 18$, $\dim C_1 = 27$, $\dim C_0 = 9$.

As usual, the image of δ_1 consists of all sums of an even number of vertices, so this has dimension $\dim C_0 - 1$, i.e., 8, so $\dim H_0 = 1$.

As $\dim \operatorname{Im} \delta_1 = 8$, we see that $\dim \operatorname{Ker} \delta_1 = 19$. Now, if we take all the 2-simplices in the torus, their boundaries will add up to 0, since each edge occurs as the face of exactly two 2-simplices. Hence this is an element of $\operatorname{Ker} \delta_2$. Moreover, an easy computation shows that it is the only non-zero 2-cycle: For if the 2-simplex σ is a summand in a given 2-cycle, then this cycle must also include every 2-simplex which shares an edge with σ. And then the cycle must include every 2-simplex which shares an edge with any of these simplices. Carrying on, we see that every 2-simplex must be in the cycle. Thus $\dim \operatorname{Ker} \delta_2 = 1$, which implies that $\dim \operatorname{Im} \delta_2 = 17$. Hence $\dim H_1 = 2$, and $\dim H_2 = 1$.

Hence $H_0 = \mathbf{Z}/2$, $H_1 = \mathbf{Z}/2 \oplus \mathbf{Z}/2$, $H_2 = \mathbf{Z}/2$, $H_i = 0$ for $i > 2$.

Example 9.5

A pair of rabbit ears (i.e., a figure of eight)

has six 1-cells and five 0-cells, so its chain complex is

$$C_1 \to C_0.$$

As usual, $\dim \mathrm{Im}\, \delta_1 = \dim C_0 - 1 = 4$, so $\dim H_0 = 1$ and $\dim H_1 = 2$. Hence $H_0 = \mathbf{Z}/2$, $H_1 = \mathbf{Z}/2 \oplus \mathbf{Z}/2$, $H_i = 0$ for $i > 1$.

Example 9.6

We can produce a simplicial sphere by taking a tetrahedron, as in Example 7.11. This has four 0-simplices, six 1-simplices (consisting of all possible pairs of vertices; note that $\binom{4}{2} = 6$) and four 2-simplices (consisting of all possible triples of vertices; $\binom{4}{3} = 4$).

As usual, $\dim H_0 = 1$, since $\mathrm{Im}\, \delta_1$ consists of all sums of two vertices, so has dimension 3. This tells us that $\dim \mathrm{Ker}\, \delta_1 = 6 - 3 = 3$. Now δ_2 is almost injective: Each 1-simplex on the boundary of a given 2-simplex is shared with only one other 2-simplex, and the only element of C_2 which is in $\mathrm{Ker}\, \delta_2$ is the sum of all four 2-simplices. Hence $\dim \mathrm{Ker}\, \delta_2 = 1$, and $\dim \mathrm{Im}\, \delta_2 = 3$.

Hence $H_0 = \mathbf{Z}/2$, $H_1 = 0$, $H_2 = \mathbf{Z}/2$ and $H_i = 0$ for $i > 2$.

Notice that in all of these examples $H_0 = \mathbf{Z}/2$. This is a particular case of the following general fact about H_0.

Proposition 9.7

For any simplicial complex K, the dimension of $H_0(K)$ is equal to the number of path components in K, i.e., $\dim H_0$ is the number of elements of $\pi_0(K)$.

Proof

The group $\pi_0(K)$ is the set of pointed maps $S^0 \to K$ modulo homotopy. Every pointed map $f : S^0 \to K$ is determined by the point $f(-1)$ in K, so $\pi_0(K)$ is equivalent to the set K modulo the relation that $x \sim y$ if, and only if, there is a continuous path $[0,1] \to K$ that takes the values x at 0 and y at 1. Each point $x \in K$ is in the interior of exactly one simplex, and is joined to each vertex of that simplex by a continuous path. So $\pi_0(K)$ is equivalent to the set S_0 of 0-simplices of K, modulo path-connectivity.

On the other hand, $H_0(K) = C_0/\text{Im } \delta_1$ is the set of $\mathbf{Z}/2$-linear combinations of vertices of K modulo the relation that two vertices, x, y are equivalent, if $y - x$ is in the image of δ_1, i.e., there is a list of 1-simplices $e_1, \ldots, e_n$ with $\delta_1(e_1 + \cdots + e_n) = y - x$. This condition means that, if we arrange $e_1, \ldots, e_n$ appropriately, e_1 is a 1-simplex from x to another vertex, x_1, e_2 is a 1-simplex from x_1 to another vertex x_2, and so on, up to e_n which is a 1-simplex from x_{n-1} to y.

If we choose a set of vertices $\{v_1, \ldots, v_m\}$ such that every vertex is equivalent to one, and only one, of these vertices under this δ_1 relation, then every element of $H_0(K)$ can be represented uniquely by a $\mathbf{Z}/2$-linear combination of $\{v_1, \ldots, v_m\}$. Hence $\dim H_0(K) = m$.

Thus a basis for $H_0(K)$ is obtained by taking the vertices of K and applying one equivalence relation, and $\pi_0(K)$ can be computed by taking the vertices of K and applying *another* equivalence relation. We will complete the proof by showing that these equivalence relations actually coincide.

If $f : [0,1] \to K$ is a path from one vertex of K to another then we can replace f by another path (in fact, one that is homotopic to f) whose image is contained in the union of the 1-simplices of K. If we take precisely those 1-simplices that form the image of this replacement path, then that forms an element of C_1 whose boundary is $f(1) - f(0)$. Hence if x and y correspond to equivalent elements of $\pi_0(K)$, then $y - x \in \text{Im } \delta_1$.

And vice versa: if $y - x \in \text{Im } \delta_1$ in H_0 then there is a list of 1-simplices whose boundary is $y - x$ and which, consequently, can be put together to form a path from x to y. Thus the two equivalence relations are the same, completing the proof of the Proposition. $\qquad\square$

Example 9.8

In the simplicial complex

there are two path components, and the chain complex for simplicial homology is

$$C_1 \to C_0$$

where $\dim C_0 = 3$, $\dim C_1 = 1$. The boundary of the single 1-simplex is nonzero, so $\dim \text{Im } \delta_1 = 1$, and so $\dim H_0 = 2$, while $\dim H_1 = 0$.

The Euler number is closely related to simplicial homology:

Proposition 9.9

If K is a simplicial complex, then

$$\chi(K) = \sum_{n \geq 0} (-1)^n \dim H_n.$$

Proof

The definition of the Euler number is $\chi(K) = \sum_{n \geq 0} (-1)^n \dim C_n(K)$. Since $H_n = \operatorname{Ker} \delta_n / \operatorname{Im} \delta_{n+1}$, its dimension is given by

$$\dim H_n = (\dim C_n - \dim \operatorname{Im} \delta_n) - \dim \operatorname{Im} \delta_{n+1}.$$

In forming the alternating sum, the last two terms on the right cancel, giving

$$\sum_{n \geq 0} (-1)^n \dim H_n = \sum_{n \geq 0} (-1)^n \dim C_n = \chi(K). \qquad \square$$

9.2 Limitations of Homology Modulo 2

All the preceding examples, the circle, square, annulus, sphere, torus and rabbit ears, have different homology except for the circle and annulus, which both have the same homology. Since the circle and annulus are homotopy equivalent, this is no great surprise. But it suggests that homology is very powerful as it can distinguish all the other examples we have computed.

Nevertheless, it is not as powerful as it could be. For example, the Klein bottle can be triangulated in a similar way to the torus, and the resulting homology groups will be the same as for the torus, with $\dim H_0 = 1$, $\dim H_1 = 2$, $\dim H_2 = 1$ and $\dim H_i = 0$ for $i > 2$. Yet the Klein bottle and the torus are not homotopy equivalent.

By looking at the torus and Klein bottle more closely we can see how to improve the homology theory that we have. To understand the similarities and differences, we imagine the triangulation of the torus as coming from a triangulation of the square pictured below. In order to get the torus from folding up this square, we would have to stretch some edges and squeeze others, but we will end up with the triangulation pictured in Example 7.8.

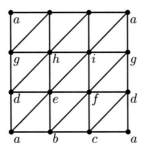

The vertex labels reflect the way that several vertices are identified with each other when we form the torus. For example, all four corners are labelled a, as they are all glued together when we form the torus. If we glue this square together to get the Klein bottle, in the way described by Example 5.57, then again, all four corners would be identified together, and the vertex labelling above would be correct for the Klein bottle. It is for this reason that the two vertices in the middle of the top row have not been labelled, as their labels change according to whether we form the torus or the Klein bottle. For the torus they should be labelled b, c, from left to right, whereas for the Klein bottle they should be labelled c, b.

If we take the chain $[a, b] + [b, c] + [c, a]$ corresponding to the bottom row, then we see that this is in Ker δ_1 as its boundary is $a + b + b + c + c + a = 0$. So this gives rise to some class in homology. And if we add anything from Im δ_2, we get the same class in homology. Well, suppose we add the boundaries of the 2-simplices $[a, e, b]$, $[b, f, c]$, $[c, d, a]$. This gives the longer chain $[a, e] + [e, b] + [b, f] + [f, c] + [c, d] + [d, a]$, corresponding to the zig-zag chain

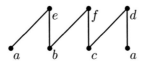

Then, adding the boundaries of the 2-simplices $[a, d, e]$, $[b, e, f]$, $[c, d, f]$, we get the chain

$$[a, d] + [d, e] + [e, b] + [e, b] + [e, f] + [f, c] + [c, f] + [f, d] + [d, a] = [d, e] + [e, f] + [f, d],$$

i.e., the second row. Carrying on in the same way, we see that $[a, b] + [b, c] + [c, a]$ gives the same homology class as the third row $[g, h] + [h, i] + [i, g]$. And, for the torus, if we move up one more row, we get back to $[a, b] + [b, c] + [c, a]$. In the Klein bottle, however, we get $[a, c] + [c, b] + [b, a]$, i.e., the same simplices, but each in the opposite direction.

As we have formulated homology, this is just the same as $[a, b] + [b, c] + [c, a]$, but if we could incorporate direction in some way, then we might be able to distinguish between these two chains and, hence, between the torus and the

Klein bottle. In particular, if we could make $[b, a]$ equal to $-[a, b]$, then, for the Klein bottle, we would have

$$
\begin{aligned}
[a, b] + [b, c] + [c, a] \equiv [a, c] + [c, b] + [b, a] &= -[c, a] - [b, c] - [a, b] \\
&= -([a, b] + [b, c] + [c, a]) \bmod \operatorname{Im} \delta_2.
\end{aligned}
$$

In other words, $2([a, b] + [b, c] + [c, a]) \in \operatorname{Im} \delta_2$, so we would have some non-zero element of $\operatorname{Im} \delta_2$, whereas in the torus, no such element arises.

Clearly, in order to distinguish $[a, b]$ from $-[a, b]$, we need to move away from $\mathbf{Z}/2$ and work, instead, with integer coefficients. We also need to establish what the analogues of $[b, a] = -[a, b]$ are for 2-simplices and higher-dimension simplices.

For a 2-simplex $[a, b, c]$, there are six different ways of ordering the vertices: $[a, b, c]$, $[b, c, a]$, $[c, a, b]$, $[b, a, c]$, $[a, c, b]$, $[c, b, a]$. Each of these can be turned into any of the others by swapping pairs of vertices repeatedly. In particular, any can be turned into $[a, b, c]$ by swapping pairs of vertices. Some require just one swap: $[b, a, c]$, $[c, b, a]$, $[a, c, b]$ while some require two swaps: $[b, c, a]$, $[c, a, b]$. With more vertices, even more swaps are necessary.

We group $[b, a, c]$, $[c, b, a]$ and $[a, c, b]$ together, being the ones which need an odd number of swaps. And we group $[a, b, c]$, $[b, c, a]$ and $[c, a, b]$ together, being those which need an even number of swaps. We then consider $[b, a, c]$, $[c, b, a]$ and $[a, c, b]$ to be the same **oriented simplex**, and we consider $[a, b, c]$, $[b, c, a]$, $[c, a, b]$ to be the same as each other, but "opposite" to $[b, a, c]$, $[c, b, a]$ and $[a, c, b]$. When working with chains of simplices, we will insist that $[b, a, c] = -[a, b, c]$.

In general, an n-simplex has $(n + 1)!$ orderings, but each can be turned into any other by a number of swaps. We consider two orderings to have the same orientation if they differ by an even number of swaps, and to have the opposite orientation if they differ by an odd number of swaps. An **oriented simplex** is then a set of orderings of the vertex list of a simplex, such that all these orderings have the same orientation, and any other ordering having the same orientation is in the set.

9.3 Integral Simplicial Homology

Given a simplicial complex K, we let $S_n(K)$ be the set of all n-simplices as before. But now we let $C_n(K)$ be the set of all $\mathbf{Z}$-linear combinations of oriented simplices, subject to the relation that if σ is an oriented simplex, then $(-1)\sigma$ is the same simplex with the opposite orientation. So, $C_n(K)$ consists of sums such as

$$
2\sigma_1 + 3\sigma_2 - 2\sigma_4,
$$

and if τ is the same simplex as σ_4 but with its vertices changed by an odd permutation, then this element of $C_n(K)$ is the same as

$$2\sigma_1 + 3\sigma_2 + 2\tau.$$

As before, an element of $C_n(K)$ is called an n-**chain**.

Now, we have seen that the boundary of a k-simplex consists of the $k+1$ different $(k-1)$-simplices obtained by omitting one of the vertices. Hence the boundary of $[v_0, v_1, v_2]$ consists of $[v_1, v_2]$, $[v_0, v_2]$ and $[v_0, v_1]$. If we imagine each of these 1-simplices as being given an arrow from the first listed vertex to the second, then the arrows on the boundary of the 2-simplex are as follows:

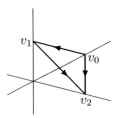

Instinctively we know that the arrow on $[v_0, v_2]$ should point the other way, so that all the arrows would be anti-clockwise. This instinct is borne out by comparing this picture with the corresponding picture for $[v_1, v_2, v_0]$. This is the same oriented simplex as $[v_0, v_1, v_2]$, yet if we orient the boundary simplices $[v_2, v_0]$, $[v_1, v_0]$, $[v_1, v_2]$ by giving each an arrow from the first vertex to the second, as above, then we get the following picture:

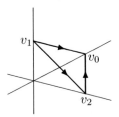

The direction of two of the arrows disagrees with the earlier picture. Since this is the boundary of the same oriented simplex, they should look the same.

The solution is to reverse the orientation on the second simplex in each case, so that the boundary of $[v_0, v_1, v_2]$ is $[v_1, v_2]$, $-[v_0, v_2]$, $[v_0, v_1]$ and the boundary of $[v_1, v_2, v_0]$ is $[v_2, v_0]$, $-[v_1, v_0]$ and $[v_1, v_2]$. Since $-[a, b] = [b, a]$, we see that these two sets of oriented simplices are the same.

Similarly, in calculating the boundary of an arbitrary simplex $[v_0, \ldots, v_n]$ we would need to negate any simplex $[v_0, \ldots, \widehat{v_i}, \ldots, v_n]$ on the boundary where i is odd. This leads us to define the boundary operator $\delta_n : C_n(K) \rightarrow C_{n-1}(K)$

by

$$\delta_n[v_0, \ldots, v_n] = \sum_{i=0}^{n}(-1)^i[v_0, \ldots, \widehat{v_i}, \ldots, v_n]$$

for a simplex $[v_0, \ldots, v_n]$, and $\delta_n(\lambda x + \mu y) = \lambda\delta_n(x) + \mu\delta_n(y)$ for any two elements $x, y \in C_n(K)$ and coefficients $\lambda, \mu \in \mathbf{Z}$. For example,

$$\delta_2[v_0, v_1, v_2] = [v_1, v_2] - [v_0, v_2] + [v_0, v_1]$$

as above, and

$$\delta_1[v_0, v_1] = v_1 - v_0.$$

Note that if we swap a pair of vertices in $[v_0, \ldots, v_n]$ then $\delta_n[v_0, \ldots, v_n]$ changes sign, i.e., δ_n respects the orientation. For example,

$$\delta_2[v_0, v_2, v_1] = [v_2, v_1] - [v_0, v_1] + [v_0, v_2] = -([v_1, v_2] + [v_2, v_0] + [v_0, v_1]).$$

From these examples, we can see that

$$\begin{aligned}(\delta_1 \circ \delta_2)[v_0, v_1, v_2] &= \delta_1[v_1, v_2] - \delta_1[v_0, v_2] + \delta_1[v_0, v_1] \\ &= (v_2 - v_1) - (v_2 - v_0) + (v_1 - v_0) = 0,\end{aligned}$$

i.e., $\delta_1 \circ \delta_2$ is zero on any 2-simplex. This generalizes to the following analogue of Lemma 9.1 which is proved in just the same way as that result.

Lemma 9.10

For every $n \geq 1$, the composite

$$\delta_n \circ \delta_{n+1} : C_{n+1} \longrightarrow C_{n-1}$$

is the zero homomorphism.

Thus we have a chain complex as before. So we define $B_n = \operatorname{Im} \delta_n$ and $Z_n = \operatorname{Ker} \delta_{n-1}$ (unless $n = 0$, in which case we set $Z_0 = C_0$), referring to elements of B_n as boundaries and elements of Z_n as cycles. The lemma assures us that $B_n \subset Z_n$, so we can define the nth **integral homology group** of the complex K to be the quotient $H_n(K) = Z_n/B_n$. For consistency, we will now use the standard terminology $H_n(K; \mathbf{Z}/2)$ to denote the **mod 2 homology** of K, i.e., the homology built around $\mathbf{Z}/2$ that we introduced in Section 9.1. To emphasize the difference, we sometimes write $H_n(K; \mathbf{Z})$ for integral homology.

Example 9.11

Let K be the simplicial circle of Example 7.1, with vertices v_0, v_1, v_2. If we choose orientations for the 1-simplices as follows: $[v_0, v_1]$, $[v_1, v_2]$, $[v_2, v_0]$, then

$$\delta_1(\lambda_0[v_0, v_1] + \lambda_1[v_1, v_2] + \lambda_2[v_2, v_0])$$
$$= \lambda_0(v_1 - v_0) + \lambda_1(v_2 - v_1) + \lambda_2(v_0 - v_2)$$
$$= (\lambda_2 - \lambda_0)v_0 + (\lambda_0 - \lambda_1)v_1 + (\lambda_1 - \lambda_2)v_2.$$

If this is 0, then $\lambda_0 = \lambda_1 = \lambda_2$. Hence Ker δ_1 consists of all integer multiples of $[v_0, v_1] + [v_1, v_2] + [v_2, v_0]$ and is isomorphic to $\mathbf{Z}$. As $\delta_2 = 0$, we see that $H_1(K) = \mathbf{Z}$.

The calculation of δ_1 above also shows that the image of δ_1 consists of all expressions $\mu_0 v_0 + \mu_1 v_1 + \mu_2 v_2$ where $\mu_2 = -(\mu_0 + \mu_1)$. Hence this image is generated by $v_0 - v_2$, $v_1 - v_2$ and is isomorphic to $\mathbf{Z}^2$. If $a_0 v_0 + a_1 v_1 + a_2 v_2$ is an arbitrary element of C_0, then we can express it as $a_0(v_0 - v_2) + a_1(v_1 - v_2) + (a_2 + a_1 + a_0)v_2$, i.e., as an element of Im δ_1 plus some multiple of v_2. Hence the quotient group C_0/B_0 is generated by v_2 and is isomorphic to $\mathbf{Z}$.

Thus $H_0(K) = \mathbf{Z}$, $H_1(K) = \mathbf{Z}$ and $H_i(K) = 0$ for $i > 1$.

Note that it is no longer appropriate to give the dimensions of the nth homology groups, because they are not vector spaces, but Abelian groups, and they need not be free. This is partly what makes integral homology more informative: It cannot be reduced to a list of dimensions in the same way as $\mathbf{Z}/2$ homology.

For most of the examples that we have met, the integral homology looks much like the $\mathbf{Z}/2$ homology (but with $\mathbf{Z}/2$ replaced by $\mathbf{Z}$):

Example 9.12

If K is the simplicial square of Example 7.2, then $H_0(K) = \mathbf{Z}$ and $H_i(K) = 0$ for $i > 0$.

Example 9.13

If K is the simplicial annulus of Example 7.3, then $H_0(K) = \mathbf{Z}$, $H_1(K) = \mathbf{Z}$ and $H_i(K) = 0$ for $i > 1$.

Example 9.14

If K is the simplicial sphere, then $H_0(K) = \mathbf{Z}$, $H_2(K) = \mathbf{Z}$ and $H_i(K) = 0$ otherwise.

Example 9.15

If K is the simplicial torus, then $H_0(K) = \mathbf{Z}$, $H_1(K) = \mathbf{Z} \oplus \mathbf{Z}$, $H_2(K) = \mathbf{Z}$ and $H_i(K) = 0$ for $i > 2$.

Note that in these examples $H_0 = \mathbf{Z}$. This is an example of the integral analogue of Proposition 9.7.

Proposition 9.16

The group $H_0(K)$ is a free Abelian group whose rank is equal to the number of path components in K.

This can be proved in exactly the same way as in the mod 2 case.

Example 9.17

If K is the Klein bottle, then we have seen that $H_0(K) = \mathbf{Z}$, as the simplicial complex is connected. Unlike the torus, $H_2(K)$ is now zero, as Ker $\delta_2 = 0$. To see why this is so, note that every 1-simplex occurs as a face of exactly two 2-simplices. For example, $[a, b]$ occurs as a face of $[a, b, e]$ and $[a, b, i]$. So if $[a, b, e]$ is a summand in a chain in Ker δ_2, then so must $[a, b, i]$ be. Moreover, the coefficient of $[a, b, e]$ determines the coefficient of $[a, b, i]$. You can quickly check that if $[a, b, i]$ is a summand in the chain, then so is $[a, i, g]$, because of their common boundary simplex $[a, i]$. And, pursuing this reasoning, you will quickly find that every 2-simplex must occur in the chain. In fact, if we order the simplices in the right way, for example anticlockwise in the earlier diagram, they must all have the same coefficient. This shows that the only possible elements of Ker δ_2 are multiples of this sum of all the 2-simplices. Then a quick calculation reveals that δ_2 is not actually zero on this sum, but is equal to $2([a, b] + [b, c] + [c, a])$. Hence Ker $\delta_2 = 0$, so $H_2(K) = 0$, in contrast to the torus.

As $H_0(K) = \mathbf{Z}$, we know that Im δ_1 is a free Abelian group of rank 8. Hence Ker δ_1 is a free Abelian group of rank $27 - 8 = 19$. Since Ker $\delta_2 = 0$, we know that Im $\delta_2 \approx C_2$ is a free Abelian group of rank 18. However, whereas the chain $[a, b] + [b, c] + [c, a]$ belongs to Ker δ_1, only $2([a, b] + [b, c] + [c, a])$ is in Im δ_2. Hence $H_1(K) = \mathbf{Z} \oplus \mathbf{Z}/2$, again differing slightly from the torus.

In all our other calculations of integral homology, the homology groups turned out to be direct sums of copies of $\mathbf{Z}$. In this example, we see a finite summand $\mathbf{Z}/2$ appearing. Such finite subgroups of integral homology groups are called **torsion** because they tend to arise from the sort of "twisting" that produces a Klein bottle instead of a torus. So, for example, we would say that the homology of a torus is "torsion-free", whereas the homology of the Klein bottle "has torsion."

The close connection between integral and $\mathbf{Z}/2$-homology in the other examples is not a coincidence, and there is a connection between the two which even explains the Klein bottle's $\mathbf{Z}/2$-homology. This is part of a general result called the universal coefficient theorem, a proof of which can be found in Section 3.A of [5].

Theorem 9.18 (Universal Coefficient Theorem for $\mathbf{Z}/2$)

For any simplicial complex K,

$$H_n(K; \mathbf{Z}/2) = (H_n(K) \otimes \mathbf{Z}/2) \oplus \mathrm{Tor}(H_{n-1}(K), \mathbf{Z}/2).$$

The **tensor product** operation $\otimes$ and the Tor operator may be unfamiliar, but it is easy to desribe their action, at least for finitely generated groups. Every finitely generated Abelian group G can be expressed as a direct sum

$$G = \mathbf{Z}^n \oplus \mathbf{Z}/p_1^{r_1} \oplus \mathbf{Z}/p_2^{r_2} \oplus \cdots \oplus \mathbf{Z}/p_m^{r_m}$$

for some non-negative integers $n, r_1, r_2, \ldots, r_m$ and primes $p_1, \ldots, p_m$. Both Tor and $\otimes$ respect this direct sum, and act on the individual summands according to the following rules:

$$\mathbf{Z} \otimes \mathbf{Z}/2 = \mathbf{Z}/2, \qquad \mathbf{Z}/2^r \otimes \mathbf{Z}/2 = \mathbf{Z}/2 \qquad \mathbf{Z}/p^r \otimes \mathbf{Z}/2, = 0 \text{ if } p \text{ is odd},$$
$$\mathrm{Tor}(\mathbf{Z}, \mathbf{Z}/2) = 0, \qquad \mathrm{Tor}(\mathbf{Z}/2^r, \mathbf{Z}/2) = \mathbf{Z}/2, \quad \mathrm{Tor}(\mathbf{Z}/p^r, \mathbf{Z}/2) = 0 \text{ if } p \text{ is odd}.$$

Thus, for example, $(\mathbf{Z} \oplus \mathbf{Z}/4 \oplus \mathbf{Z}/3) \otimes \mathbf{Z}/2 = \mathbf{Z}/2 \oplus \mathbf{Z}/2$, whereas $\mathrm{Tor}(\mathbf{Z} \oplus \mathbf{Z}/4 \oplus \mathbf{Z}/3, \mathbf{Z}/2) = \mathbf{Z}/2$.

Example 9.19

Given that the integral homology of the Klein bottle is

$$H_i(K) = \begin{cases} \mathbf{Z} & \text{if } i = 0, \\ \mathbf{Z} \oplus \mathbf{Z}/2 & \text{if } i = 1, \\ 0 & \text{if } i > 1, \end{cases}$$

the universal coefficient theorem then tells us that

$$H_0(K; \mathbf{Z}/2) = \mathbf{Z} \otimes \mathbf{Z}/2 = \mathbf{Z}/2,$$
$$H_1(K; \mathbf{Z}/2) = (\mathbf{Z} \oplus \mathbf{Z}/2) \otimes \mathbf{Z}/2 \oplus \mathrm{Tor}(\mathbf{Z}, \mathbf{Z}/2) = \mathbf{Z}/2 \oplus \mathbf{Z}/2,$$
$$H_2(K; \mathbf{Z}/2) = (0 \otimes \mathbf{Z}/2) \oplus \mathrm{Tor}(\mathbf{Z} \oplus \mathbf{Z}/2, \mathbf{Z}/2) = \mathbf{Z}/2,$$
$$H_i(K; \mathbf{Z}/2) = (0 \otimes \mathbf{Z}/2) \oplus \mathrm{Tor}(0, \mathbf{Z}/2) = 0 \quad \text{if } i > 2,$$

which agrees with our earlier calculation.

The universal coefficient theorem shows that integral homology contains all the information that $\mathbf{Z}/2$ homology contains, while the Klein bottle shows that it actually contains *more* information, as only integral homology can distinguish the Klein bottle from the torus.

EXERCISES

9.1. Calculate the mod 2 homology of the simplicial annulus of Example 7.3.

9.2. Triangulate the closed interval $[0, 1]$ and calculate its mod 2 homology. Verify that the result does not change if you use a different triangulation.

9.3. Triangulate the cylinder $S^1 \times [0, 1]$ and calculate its mod 2 homology and its integral homology.

9.4. Triangulate the Möbius band and calculate its mod 2 homology and its integral homology. Verify that the universal coefficient theorem holds for this space.

9.5. Calculate the integral homology of the simplicial annulus, of the simplicial square of Example 7.2, and of the simplicial sphere of Example 7.11, and verify the results given in Examples 9.12, 9.13 and 9.14.

9.6. Take the simplicial torus of Example 7.8, and glue in another 2-simplex joining the three innermost edges. How does gluing in this 2-simplex change the homology? Compare the change in the homology with the change in the Euler characteristic.

10
Singular Homology

For simplicial homology we supposed that our space had already been expressed as a simplicial complex, i.e., decomposed into a union of simplices. Some spaces cannot be expressed in such a way, and even those that can, can usually be expressed as a simplicial complex in many different ways. Choosing one way can obscure some details of the space. For these reasons "singular" homology was developed, which gets around this problem by, in a very loose sense, considering all possible simplicial decompositions.

10.1 Singular Homology

Define the **standard n-simplex** $\Delta^n \subset \mathbf{R}^{n+1}$ to be the n-simplex with vertices $(1,0,\ldots,0)$, $(0,1,0,\ldots,0)$, $\ldots$, $(0,\ldots,0,1)$, i.e.,

$$\Delta^n = \{(x_0,\ldots,x_n) \in \mathbf{R}^{n+1} : x_i \geq 0 \text{ for all } i; \sum_{i=0}^{n} x_i = 1\}.$$

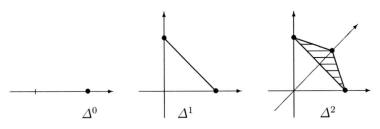

Δ^0 Δ^1 Δ^2

M.D. Crossley, *Essential Topology*, Springer Undergraduate
Mathematics Series, DOI 10.1007/978-1-84628-194-5_10,
© Springer-Verlag London Limited 2010

In a simplicial complex K, an n-simplex can be thought of as the image of a continuous injection $\Delta^n \to K$. A triangulation of a space X can then be thought of as a list of continuous injections from standard simplices to X. Rather than choosing such a list we now consider all continuous maps from standard simplices to X, non-injective as well as injective. Such a map need not preserve the topology of Δ^n very well. For example, the following picture shows the images of three different maps $\Delta^1 \to \mathbf{R}^2$, none of which is homeomorphic with Δ^1.

Since the image of f bears little relation to Δ^n, such a map is called a **singular n-simplex** and we write $S_n(X)$ for the set of all singular n-simplices $\Delta^n \to X$, abbreviating this to S_n if the context is sufficiently clear.

As with simplicial homology, we need some information about how these simplices join together to form the space, and we do this by looking at the boundary of each simplex.

The boundary of an n-simplex is the union of its faces, and for Δ^n these have a particularly simple description: If we define $d^i : \mathbf{R}^n \to \mathbf{R}^{n+1}$, for $0 \le i \le n$, to be the map

$$d^i(x_0, \ldots, x_{n-1}) = (x_0, \ldots, x_{i-1}, 0, x_i, \ldots, x_{n-1}),$$

then $d^i(\Delta^{n-1})$ is contained in Δ^n, so d^i restricts to a map $\Delta^{n-1} \to \Delta^n$. Moreover, $d^i(\Delta^{n-1})$ is a face of Δ^n and the boundary of Δ^n is

$$d^0(\Delta^{n-1}) \cup d^1(\Delta^{n-1}) \cup \cdots \cup d^n(\Delta^{n-1})$$

i.e., the set of all points in Δ^n that have at least one coordinate equal to 0.

Consequently, if we have a singular n-simplex $f : \Delta^n \to X$, then we can compose with the $n + 1$ different maps d^i, to get $n + 1$ different maps $f \circ d^i : \Delta^{n-1} \to X$, i.e., $n+1$ different singular $(n-1)$-simplices. So the boundary of a singular n-simplex is a collection of $n + 1$ singular $(n-1)$-simplices, just as for the simplices in a simplicial complex. This means that we can define a boundary operator just as we did for simplicial homology. We define $C_n(X)$ (or just C_n) to be the free Abelian group on $S_n(X)$, i.e., the set of $\mathbf{Z}$-linear combinations of singular n-simplices.[1] Elements of C_n are called **singular n-chains** and the boundary of a singular n-simplex f is the singular $(n - 1)$-chain

$$\delta_n(f) = f \circ d^0 - f \circ d^1 + f \circ d^2 - \cdots \pm f \circ d^n = \sum_{i=0}^{n} (-1)^i f \circ d^i.$$

---

[1] Note that we do not need to worry about orientation, as that is taken care of by considering a simplex as a continuous map and not just its image.

Extending this additively gives us a homomorphism $\delta_n : C_n \to C_{n-1}$, called the **singular boundary operator**. For example, if $f : \Delta^2 \to X$ is a 2-simplex, then $\delta_2(f)$ is the 1-chain given by

$$\delta_2(f)(x,y) = f(d^0(x,y)) - f(d^1(x,y)) + f(d^2(x,y))$$
$$= f(0,x,y) - f(x,0,y) + f(x,y,0),$$

and if $f : \Delta^3 \to X$ is a 3-simplex, then $\delta_2(\delta_3(f))$ is the 1-chain given by

$$\delta_2(\delta_3(f))(x,y)$$
$$= \delta_3(f)(d^0(x,y)) - \delta_3(f)(d^1(x,y)) + \delta_3(f)(d^2(x,y))$$
$$= \delta_3(f)(0,x,y) - \delta_3(f)(x,0,y) + \delta_3(f)(x,y,0)$$
$$= (f(d^0(0,x,y)) - f(d^1(0,x,y)) + f(d^2(0,x,y)) - f(d^3(0,x,y)))$$
$$\quad -(f(d^0(x,0,y)) - f(d^1(x,0,y)) + f(d^2(x,0,y)) - f(d^3(x,0,y)))$$
$$\quad +(f(d^0(x,y,0)) - f(d^1(x,y,0)) + f(d^2(x,y,0)) - f(d^3(x,y,0)))$$
$$= f(0,0,x,y) - f(0,0,x,y) + f(0,x,0,y) - f(0,x,y,0)$$
$$\quad -f(0,x,0,y) + f(x,0,0,y) - f(x,0,0,y) + f(x,0,y,0)$$
$$\quad +f(0,x,y,0) - f(x,0,y,0) + f(x,y,0,0) - f(x,y,0,0);$$

and, as you can see, all the terms cancel, so that $\delta_2(\delta_3(f)) = 0$. This generalizes to the following analogue of Lemma 9.1:

Lemma 10.1

For every $n \geq 1$, the composite

$$\delta_n \circ \delta_{n+1} : C_{n+1} \longrightarrow C_{n-1}$$

is the zero homomorphism.

Proof

Note first that the composite $d^i \circ d^j : \Delta^{n-1} \to \Delta^{n+1}$ satisfies

$$d^i \circ d^j(x_0, \ldots, x_{n-1}) = d^i(x_0, \ldots, x_{j-1}, 0, x_j, \ldots, x_{n-1})$$
$$= \begin{cases} (x_0, \ldots, x_{j-1}, 0, x_j, \ldots, x_{i-2}, 0, x_{i-1}, \ldots, x_{n-1}) & \text{if } i > j, \\ (x_0, \ldots, x_{i-1}, 0, x_i, \ldots, x_{j-1}, 0, x_j, \ldots, x_{n-1}) & \text{if } i \leq j. \end{cases}$$

From this we can see that

$$d^i \circ d^j = \begin{cases} d^j \circ d^{i-1} & \text{if } i > j, \\ d^{j+1} \circ d^i & \text{if } i \leq j. \end{cases}$$

As in the simplicial case, Lemma 9.1, because δ_n and δ_{n+1} are both additive functions, it is enough to check that $\delta_n \circ \delta_{n+1}(f) = 0$ for an $(n+1)$-simplex $f : \Delta^{n+1} \to X$. For such a simplex, we have

$$
\delta_n \delta_{n+1}(f) = \sum_{j=0}^{n}(-1)^j \left(\sum_{i=0}^{n+1}(-1)^i f \circ d^i \right) \circ d^j = \sum_{j=0}^{n}\sum_{i=0}^{n+1}(-1)^{i+j} f \circ (d^i \circ d^j)
$$

$$
= \sum_{j=0}^{n}\sum_{i=0}^{j}(-1)^{i+j} f \circ (d^i \circ d^j) + \sum_{j=0}^{n}\sum_{i=j+1}^{n+1}(-1)^{i+j} f \circ (d^i \circ d^j)
$$

$$
= \sum_{j=0}^{n}\sum_{i=0}^{j}(-1)^{i+j} f \circ (d^i \circ d^j) + \sum_{j=0}^{n}\sum_{i=j+1}^{n+1}(-1)^{i+j} f \circ (d^j \circ d^{i-1})
$$

$$
= \sum_{j=0}^{n}\sum_{i=0}^{j}(-1)^{i+j} f \circ (d^i \circ d^j) + \sum_{i=0}^{n}\sum_{j=i+1}^{n+1}(-1)^{j+i} f \circ (d^i \circ d^{j-1})
$$

$$
= \sum_{j=0}^{n}\sum_{i=0}^{j}(-1)^{i+j} f \circ (d^i \circ d^j) + \sum_{j=0}^{n}\sum_{i=0}^{j}(-1)^{i+j+1} f \circ (d^i \circ d^j)
$$

$$
= \sum_{j=0}^{n}\sum_{i=0}^{j}(-1)^{i+j} \left(f \circ (d^i \circ d^j) - f \circ (d^i \circ d^j) \right) = 0. \qquad \square
$$

The collection $C_*(X) = \{C_n, \delta_n\}$ of groups and boundary operators is thus a chain complex, and is called the **singular chain complex** for X. The remainder of the construction of singular homology copies that of simplicial homology: We define $Z_n(X) = \text{Ker } \delta_n$ for $n > 0$ and $Z_0(X) = C_0(X)$, calling elements of $Z_n(X)$ **cycles**, and we set $B_n(X) = \text{Im } \delta_{n+1}$, elements of $B_n(X)$ being **boundaries**. Then we define singular homology as follows.

Definition: The nth **singular homology group** of a topological space X is the quotient group

$$
H_n(X) = \frac{Z_n(X)}{B_n(X)} = \begin{cases} \text{Ker } \delta_n/\text{Im } \delta_{n+1} & \text{if } n > 0, \\ C_0/\text{Im } \delta_1 & \text{if } n = 0. \end{cases}
$$

The **singular homology** of X is the collection

$$
H_*(X) = \{H_0(X), H_1(X), H_2(X), \cdots\}.
$$

Example 10.2

If X consists of a single point, then we can calculate $H_*(X)$ as follows.

For each $n \geq 0$, there is just a single function $\Delta^n \to X$, as X is just a single point. Therefore, S_n contains just one element and $C_n = \mathbf{Z}$. If we take a simplex $f : \Delta^n \to X$, then $f \circ d^i : \Delta^{n-1} \to X$ is the same, no matter what i is. Hence

$$\delta_n(f) = \sum_{i=0}^{n}(-1)^i(f \circ d^i) = \sum_{i=0}^{n}(-1)^i(f \circ d^0) = \begin{cases} 0 & \text{if } n \text{ is odd,} \\ f \circ d^0 & \text{if } n \text{ is even.} \end{cases}$$

Thus $\delta_n : C_n \to C_{n-1}$ is the identity when n is even, and zero when n is odd. Hence Ker δ_n is $\mathbf{Z}$ if n odd, and 0 if n even, exactly matching Im δ_{n+1}. Hence $H_i(X) = 0$ for $i > 0$, i.e., X is acyclic, and $H_0(X) = C_0/\text{Im } \delta_1 = \mathbf{Z}/0 = \mathbf{Z}$.

Example 10.3

The same method can be used if X is a discrete space with a finite number, m, of points. Since Δ^n is connected, the image of $f : \Delta^n \to X$ must be a single point, so there are m singular n-simplices for each $n \geq 0$, and C_n is a free Abelian group of rank m. Given a simplex f, $f \circ d^i$ is independent of i, so δ_n is either 0 or an isomorphism, according to whether n is odd or even. Consequently $H_i(X) = 0$ for $i > 0$, and $H_0(X)$ is a free Abelian group of rank m. This also follows from Proposition 10.4.

As with simplicial homology, H_0 counts the number of path components, thus giving exactly the same information as π_0 (more precisely: $H_0(X)$ is isomorphic to the free Abelian group $\mathbf{Z}[\pi_0(X)]$ on $\pi_0(X)$). However, unlike the higher homotopy groups, the singular homology groups give information about all the path components of a space, not just one.

Proposition 10.4

If a space X has path components $\{P_j\}$ indexed by j belonging to some set J, then for each $i \geq 0$, $H_i(X)$ is the direct sum of the groups $H_i(P_j)$.

Proof

Because Δ^n is path connected, a singular n-simplex of X is a singular n-simplex of P_j for some $j \in J$. A singular n-chain s in X can then be expressed as a sum

$$\sum_{j \in J} s_j$$

where each s_j is a singular n-chain in P_j. This expresses $C_n(X)$ as a direct sum $\oplus_{j \in J} C_n(P_j)$. The boundary operator respects this splitting since if the image of $f : \Delta^n \to X$ is contained in P_j, then so is the image of $f \circ d^i$ for any i. Hence $H_n(X) = \oplus_{j \in J} H_n(P_j)$, as claimed. □

Corollary 10.5

If a space X has n path components, then $H_0(X)$ is a free Abelian group of rank n.

Corollary 10.6

If a space X can be written as a disjoint union $X = U \amalg V$, then, for each $i \geq 0$,

$$H_i(X) = H_i(U) \oplus H_i(V).$$

As well as the discrete spaces we considered in Examples 10.2 and 10.3, we can also calculate the singular homology of a disc.

Example 10.7

The closed n-disc D^n is acyclic, i.e., $H_k(D^n) = 0$ if $k > 0$, and, as Corollary 10.5 shows, $H_0(D^n) = \mathbf{Z}$.

The calculation of $H_k(D^n) = 0$ for $k > 0$ relies heavily on the fact that D^n is convex, i.e., given any two points $a, b \in D^n$, every point on the straight line from a to b is also contained in D^n.

Let x be any point in D^n. Given a k-simplex $f : \Delta^k \to D^n$, where $k \geq 1$, we use x to define a $(k+1)$-simplex $C_x(f)$ as follows.

The map d^{k+1} gives an inclusion $\Delta^k \to \Delta^{k+1}$, and we define $C_x(f)$ on the image of this so as to agree with f on Δ^k, i.e., so that $C_x(f) \circ d^{k+1} = f$.

The image of d^{k+1} includes every vertex of Δ^{k+1} except one, namely $(0, \ldots, 0, 1)$. If we consider all lines from this extra vertex $(0, \ldots, 0, 1)$ to some point of $d^{k+1}(\Delta^k)$, we see that every point of Δ^k lies on such a line. This can be seen for Δ^2 in the picture.

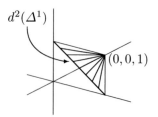

The function $C_x(f)$ is defined on every point of $d^{k+1}(\Delta^k)$, and we define $C_x(f)(0, \ldots, 0, 1)$ to be x. We can then define $C_x(f)$ on every point of a line from $(0, \ldots, 0, 1)$ to a point in $d^{k+1}(\Delta^k)$ by interpolating between the values of $C_x(f)$ on the endpoints. In other words, define

$$C_x(f)\left(t(0, \ldots, 0, 1) + (1-t)(x_0, \ldots, x_k, 0)\right) = tx + (1-t)f(x_0, \ldots, x_k)$$

for $0 \leq t \leq 1$. The right-hand side belongs to D^n because D^n is convex. Moreover, if f is continuous, then $C_x(f)$ will also be continuous and hence $C_x(f)$ is a $(k+1)$-simplex. By extending this additively, we obtain a homomorphism $C_x : C_k(D^n) \rightarrow C_{k+1}(D^n)$.

From the construction we can calculate $\delta_{k+1}(C_x(f))$ directly, and we find that

$$\delta_{k+1}(C_x(f)) = C_x(\delta_k(f)) + (-1)^{k+1}f.$$

Hence, if $f \in Z_k$, then $f = \pm\delta_{k+1}(C_x(f))$, since $\delta_k(f) = 0$. In other words, if $f \in Z_k$, then $f \in B_k$, i.e., $H_k = Z_k/B_k = 0$ for all $k > 0$.

The same argument can be applied to any convex set:

Proposition 10.8

If $X \subset \mathbf{R}^n$ is any convex subspace, then X is acyclic and $H_0(X) = \mathbf{Z}$ (since every convex subset is path connected).

10.2 Homology and Continuous Maps

We will now show that the singular homology groups behave a little like homotopy groups in that they assign an induced homomorphism to every continuous map.

Theorem 10.9

Given a continuous map $f : X \rightarrow Y$, there is an **induced homomorphism** $f_* : H_n(X) \rightarrow H_n(Y)$ for each $n \geq 0$, with the following properties:

1. If $g : Y \rightarrow Z$ is another continuous map, then $(g \circ f)_* = g_* \circ f_*$.

2. If $i : X \rightarrow X$ is the identity map, then i_* is the identity homomorphism $H_n(X) \rightarrow H_n(X)$ for each n.

3. If $h : X \rightarrow Y$ is homotopic to f, then $h_* = f_*$.

Proof

As with homotopy groups, the construction is based on composition. If we take a singular n-simplex in X, i.e., a continuous map $s : \Delta^n \rightarrow X$, then we can compose it with f to get a continuous map $f \circ s : \Delta^n \rightarrow Y$, i.e., a singular

n-simplex in Y. Thus f gives a function $S_n(X) \to S_n(Y)$ and, hence, a group homomorphism $C_n(f) : C_n(X) \to C_n(Y)$, for each $n \geq 0$. Moreover, this homomorphism respects the boundary operator in the sense that the following diagram

$$
\begin{array}{ccc}
C_n(X) & \xrightarrow{\ C_n(f)\ } & C_n(Y) \\
\delta_n \downarrow & & \downarrow \delta_n \\
C_{n-1}(X) & \xrightarrow{\ C_{n-1}(f)\ } & C_{n-1}(Y)
\end{array}
$$

is commutative, i.e., $\delta_n \circ C_n(f) = C_{n-1}(f) \circ \delta_n$. To see this, note that, for an n-simplex s, we have

$$
\delta_n \circ C_n(f)(s) = \delta_n(f \circ s) = \sum_{i=0}^{n}(-1)^i(f \circ s) \circ d^i = \sum_{i=0}^{n}(-1)^i f \circ (s \circ d^i)
$$

$$
= C_{n-1}(f) \circ \sum_{i=0}^{n}(-1)^i(s \circ d^i) = C_{n-1}(f) \circ \delta_n(s).
$$

If this holds for an n-simplex, then it holds for any element of $C_n(X)$, since both $C_n(f)$ and δ_n are group homomorphisms. In particular, for any $n \geq 1$, if $x \in \mathrm{Ker}\, \delta_n$, then $C_n(f)(x) \in \mathrm{Ker}\, \delta_n$, and if $x \in \mathrm{Im}\, \delta_n$, then $C_{n-1}(f)(x) \in \mathrm{Im}\, \delta_n$. So the function $C_n(f)$ leads to a homomorphism $f_* : H_n(X) \to H_n(Y)$, for each $n \geq 0$.

This is a special case of a more general phenomenon: If we have two chain complexes (C_*, δ_*), (D_*, ∂_*), then we say that a map $f_* = \{f_n : C_n \to D_n : n \geq 0\}$ is a **chain map** if the following diagram commutes:

$$
\begin{array}{ccc}
C_n & \xrightarrow{\ f_n\ } & D_n \\
\delta_n \downarrow & & \downarrow \partial_n \\
C_{n-1} & \xrightarrow{\ f_{n-1}\ } & D_{n-1}
\end{array}
$$

i.e., $\partial_n \circ f_n = f_{n-1} \circ \delta_n$, for all $n \geq 1$. The argument above shows that chain maps induce homomorphisms on homology.

Returning to the particular case where the chain map arises from a continuous map between topological spaces, properties 1 and 2 follow directly from the fact that the construction is based on composition.

Property 3, that homology respects homotopies, is much, much harder to prove, and the next section is devoted to proving it.   □

Before we turn to the proof of part 3, we note the following consequence of the theorem, that follows just as the analogous statement did for homotopy groups (Proposition 8.15).

Corollary 10.10

If X and Y are homotopy equivalent, then, for each $n \geq 0$, $H_n(X)$ and $H_n(Y)$ are isomorphic groups.

10.3 Homology Respects Homotopies

Theorem 10.11

If $f : X \to Y$ and $g : X \to Y$ are homotopic continuous maps, then $f_* = g_* : H_*(X) \to H_*(Y)$.

    To prove Theorem 10.11, we will show that a homotopy between two continuous maps leads to a "chain homotopy" at the level of chain complexes which, in turn, implies the equality of the induced maps. We will begin by looking at this notion of chain homotopy.

Definition: Let (C_*, δ_*), (D_*, ∂_*) be two chain complexes, and $\phi, \psi : C_* \to D_*$ two chain maps. A family of homomorphisms

$$\Phi_* = \{\Phi_n : C_n \to D_{n+1}\}_{n \geq 0}$$

is called a **chain homotopy** between ϕ and ψ if

$$\partial_{n+1}\Phi_n + \Phi_{n-1}\delta_n = \psi_n - \phi_n$$

for all $n > 0$, and

$$\partial_1 \Phi_0 = \psi_0 - \phi_0.$$

    This looks strange at first, but it is justified by the following result.

Proposition 10.12

If $\phi, \psi : (C_*, \delta_*) \to (D_*, \partial_*)$ are chain maps and Φ_* is a chain homotopy between ϕ and ψ, then the maps induced on homology by ϕ and ψ are equal.

Proof

Suppose that $s \in C_0$. Then

$$\psi_0(s) - \phi_0(s) = \partial_1 \Phi_0(s) \in \operatorname{Im} \partial_1.$$

Hence, as far as homology is concerned, $\psi_0(s) = \phi_0(s)$.

Now let $n > 0$, and suppose that $s \in C_n$ is in Ker δ_n. Then

$$\psi_n(s) - \phi_n(s) = \partial_{n+1}\Phi_n(s) + \Phi_{n-1}\delta_n(s) = \partial_{n+1}\Phi_n(s) + 0$$

since $\delta_n(s) = 0$. Hence $\psi_n(s) - \phi_n(s)$ is, again, in the image of ∂_{n+1}. So $\psi_n(s)$ and $\phi_n(s)$ are in the same homology class. As this is true for all s, the induced maps are equal.   □

Proof (of Theorem 10.11)

If we have a homotopy $F : X \times I \to Y$ such that $F(x,0) = f(x)$, $F(x,1) = g(x)$ for all $x \in X$, and we can construct a chain homotopy from F, between $C_*(f)$ and $C_*(g)$, then the preceding proposition will finish the proof, showing that $f_* = g_*$.

Such a chain homotopy would be a function $C_n(X) \to C_{n+1}(Y)$. Now, we can use $C_{n+1}(F)$ to get from $C_{n+1}(X \times I)$ to $C_{n+1}(Y)$, so we need some way of mapping from $C_n(X)$ to $C_{n+1}(X \times I)$. We will do that by constructing an element $\alpha_n \in C_{n+1}(\Delta^n \times I)$. This is just a linear combination of continuous maps $\Delta^{n+1} \to \Delta^n \times I$ and, given a simplex $s : \Delta^n \to X$ in $C_n(X)$, we can form the composite

$$\Delta^{n+1} \longrightarrow \Delta^n \times I \xrightarrow{s \times id} X \times I,$$

of $s \times id$ (where id denotes the identity map of I) with each summand of α_n. By taking the corresponding linear combination of these composites, we get an element of $C_{n+1}(X \times I)$ which, by abuse of notation, we write as $(s \times id) \circ \alpha_n$. Thus α_n gives a way of turning an n-simplex for X into an $(n + 1)$-chain for $X \times I$. Extending this additively, we get a homomorphism $C_n(X) \to C_{n+1}(X \times I)$.

Now, of course, we need to choose α_n carefully, so that the function $C_n(X) \to C_{n+1}(Y)$ that we get by this process will be a chain homotopy. Nevertheless, there are many ways of doing this, and the choices involved make it difficult to see what is going on. So, rather than write down an explicit map α_n, we will use an inductive method to construct a map with the required properties. The resulting map will be a "black box," whose details we will not understand, except that it will have the necessary properties in order to give a chain homotopy.

To be precise, we will construct $\alpha_n \in C_{n+1}(\Delta^n \times I)$ so that the following two equations are satisfied:

$$\delta_1(\alpha_0) = G_0$$

and

$$\delta_{n+1}(\alpha_n) + \sum_{j=0}^{n}(-1)^j(d^j \times id) \circ \alpha_{n-1} = G_n \quad \text{for all } n > 0,$$

where $G_n \in C_n(\Delta^n \times I)$ is the chain $G_n = e_1 - e_0$, the simplices $e_0, e_1 : \Delta^n \to \Delta^n \times I$ being defined by $e_0(\mathbf{x}) = (\mathbf{x}, 0)$ and $e_1(\mathbf{x}) = (\mathbf{x}, 1)$. The map id is the identity on I, as before.

We will first describe the construction of α_n and later show how these two equations ensure that α_n leads to a chain homotopy between $C_*(f)$ and $C_*(g)$.

To begin with, we will define α_0 by

$$\alpha_0(x_1, x_2) = (1, x_2) \in \Delta^0 \times I,$$

for all $(x_1, x_2) \in \Delta^1$. Note that if $(x_1, x_2) \in \Delta^1$, then $0 \le x_2 \le 1$, so $(1, x_2)$ is in $\Delta^0 \times I$ as required. To calculate $\delta_1(\alpha_0)$, note that this is a 0-simplex, so we only need to know $\delta_1(\alpha_0)(1)$, which is given by

$$\delta_1(\alpha_0)(1) = \alpha_0(0, 1) - \alpha_0(1, 0) = (1, 1) - (1, 0) = G_0(1).$$

Hence $\delta_1(\alpha_0) = G_0$, as required.

Next we need to construct $\alpha_1 \in C_2(\Delta^1 \times I)$ such that

$$\delta_2(\alpha_1) + \sum_{j=0}^{1} (-1)^j (d^j \times id) \circ \alpha_0 = G_1.$$

Although it seems strange, we will do this by first calculating

$$
\begin{aligned}
\delta_1 \left(\sum_{j=0}^{1} (-1)^j (d^j \times id) \circ \alpha_0 \right) &= \sum_{k=0}^{1} (-1)^k \left(\sum_{j=0}^{1} (-1)^j (d^j \times id) \circ \alpha_0 \right) \circ d^k \\
&= \sum_{j=0}^{1} (-1)^j (d^j \times id) \circ \sum_{k=0}^{1} (-1)^k \alpha_0 \circ d^k \\
&= \sum_{j=0}^{1} (-1)^j (d^j \times id) \circ \delta_1(\alpha_0) \\
&= \sum_{j=0}^{1} (-1)^j (d^j \times id) \circ G_0.
\end{aligned}
$$

This is a 0-chain, and when we apply it to the element $1 \in \Delta^0$, we get

$$\sum_{j=0}^{1} (-1)^j (d^j \times id) \circ G_0(1) = (0, 1, 1) - (1, 0, 1) - (0, 1, 0) + (1, 0, 0).$$

Similarly, if we take $\delta_1(G_1)$, and apply it to $1 \in \Delta^0$, we also get

$$\delta_1(G_1)(1) = (0, 1, 1) - (1, 0, 1) - (0, 1, 0) + (1, 0, 0).$$

In other words,

$$\delta_1 \left(G_1 - \sum_{j=0}^{1} (-1)^j (d^j \times id) \circ \alpha_0 \right) = 0,$$

i.e., we have an element of Ker δ_1. Now $\Delta^1 \times I$ is convex so, by Proposition 10.8, its homology is trivial in positive degrees, and Ker $\delta_1 = \text{Im } \delta_2$. In other words, there must be an element, which we call α_1, such that

$$G_1 - \sum_{j=0}^{1} (-1)^j (d^j \times id) \circ \alpha_0 = \delta_2(\alpha_1).$$

Rearranging this gives exactly the equation that α_1 is required to satisfy. Hence, without writing down a specific expression for it, we have found an element α_1 with the necessary properties.

Now we can proceed to construct α_n inductively for all $n > 1$ in a similar way. We suppose that $\alpha_{n-1} \in C_n(\Delta^{n-1} \times I)$ has been constructed such that

$$\delta_n(\alpha_{n-1}) + \sum_{j=0}^{n-1} (-1)^j (d^j \times id) \circ \alpha_{n-2} = G_{n-1}.$$

Then

$$\delta_n \left(\sum_{i=0}^{n} (-1)^i (d^i \times id) \circ \alpha_{n-1} \right) = \sum_{k=0}^{n} (-1)^k \left(\sum_{i=0}^{n} (-1)^i (d^i \times id) \circ \alpha_{n-1} \right) \circ d^k$$

$$= \sum_{i=0}^{n} (-1)^i (d^i \times id) \circ \delta_n(\alpha_{n-1})$$

$$= \sum_{i=0}^{n} (-1)^i (d^i \times id) \circ \left(G_{n-1} - \sum_{j=0}^{n-1} (-1)^j (d^j \times id) \circ \alpha_{n-2} \right)$$

$$= \sum_{i=0}^{n} (-1)^i (d^i \times id) \circ G_{n-1} - \sum_{i=0}^{n} \sum_{j=0}^{n-1} (-1)^{i+j} (d^i \times id) \circ (d^j \times id) \circ \alpha_{n-2}$$

$$= \sum_{i=0}^{n} (-1)^i (d^i \times id) \circ G_{n-1} - \sum_{i=0}^{n} \sum_{j=0}^{n-1} (-1)^{i+j} ((d^i \circ d^j) \times id) \circ \alpha_{n-2}$$

$$= \sum_{i=0}^{n} (-1)^i (d^i \times id) \circ G_{n-1},$$

the second term being zero because of cancellation exactly as in the proof of Lemma 10.1.

Now

$$\sum_{i=0}^{n} (-1)^i (d^i \times id) \circ G_{n-1} = \sum_{i=0}^{n} (-1)^i (e_1 - e_0) \circ d^i = \delta_n(G_n).$$

In other words,

$$G_n - \sum_{i=0}^{n}(-1)^i(d^i \times id) \circ \alpha_{n-1} \in \mathrm{Ker}\ \delta_n$$

and so, as $\Delta^n \times I$ is convex, there is an element $\alpha_n \in C_{n+1}(\Delta^n \times I)$ such that

$$G_n - \sum_{i=0}^{n}(-1)^i(d^i \times id) \circ \alpha_{n-1} = \delta_{n+1}(\alpha_n).$$

This completes the proof of the inductive step and thus we can conclude that there are elements $\alpha_n \in C_{n+1}(\Delta^n \times I)$ for all $n \geq 0$, with the required property.

As explained earlier, we use α_n to construct a homomorphism $C_n(X) \to C_{n+1}(Y)$ by sending an n-simplex $s : \Delta^n \to X$ to

$$\Phi_n(s) = F \circ (s \times id) \circ \alpha_n \in C_{n+1}(Y),$$

and extending additively. To check that this is indeed a chain homotopy between f and g, we calculate:

$$(\delta_{n+1}\Phi_n + \Phi_{n-1}\delta_n)(s) = \delta_{n+1}(F \circ (s \times id) \circ \alpha_n) + F \circ (\delta_n(s) \times id) \circ \alpha_{n-1}$$

$$= \sum_{j=0}^{n+1}(-1)^j(F \circ (s \times id) \circ \alpha_n \circ d^j) + \sum_{j=0}^{n}(-1)^j F \circ ((s \circ d^j) \times id) \circ \alpha_{n-1}$$

$$= F \circ (s \times id) \circ \left(\sum_{j=0}^{n+1}(-1)^j \alpha_n \circ d^j + \sum_{j=0}^{n}(-1)^j(d^j \times id) \circ \alpha_{n-1} \right)$$

$$= F \circ (s \times id) \circ \left(\delta_{n+1}(\alpha_n) + \sum_{j=0}^{n}(-1)^j(d^j \times id) \circ \alpha_{n-1} \right)$$

$$= F \circ (s \times id) \circ G_n.$$

Hence, applied to an element $\mathbf{x} \in \Delta^n$, this gives $F(s\mathbf{x}, 1) - F(s\mathbf{x}, 0) = g(s\mathbf{x}) - f(s\mathbf{x})$. In other words,

$$(\delta_{n+1}\Phi_n + \Phi_{n-1}\delta_n)(s) = g(s) - f(s),$$

i.e.,

$$(\delta_{n+1}\Phi_n + \Phi_{n-1}\delta_n) = C_n(g) - C_n(f),$$

as required. So Φ is a chain homotopy between f and g and, consequently, these have the same induced map in homology. $\qquad\square$

10.4 Barycentric Subdivision

In order to make it easier to compute singular homology groups, we would like to have some analogue of the Van Kampen theorem, so as to describe the homology of a space X in terms of subsets U and V which cover X. To do this, we need to be able to take a chain in $C_n(X)$ and express it as a sum of chains in $C_n(U)$ and $C_n(V)$. The domain splitting proposition, 6.29 suggests that we should be able to do this if we can restrict to sufficiently small parts of the domain, Δ^n. So we need to have some way of splitting Δ^n into smaller copies of itself in a way that will be respected by homology. We call such a splitting a **subdivision** of Δ^n. For example, we can subdivide Δ^1 by splitting it in half, into two smaller copies of itself:

For Δ^2, we subdivide into six smaller triangles:

We can think of this in terms of the subdivision of Δ^1 as follows: Each boundary edge of Δ^2 is homeomorphic to Δ^1, and we subdivide each edge in the same way as we did for Δ^1. We then add an extra vertex in the centre of Δ^2, called the **barycentre**, and join this barycentric vertex to each simplex of the subdivided edges:

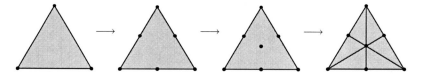

Similarly, we subdivide Δ^n inductively: If we know how to subdivide Δ^{n-1}, then we can subdivide each face of Δ^n, and we add a **barycentric vertex** in the centre of Δ^n, joining this to each simplex of the subdivided faces. This gives a simplicial complex homeomorphic with Δ^n, and each simplex in this complex has a smaller diameter than that of Δ^n. In fact, the diameter of each simplex will be at most $n/(n+1)$ times the diamater, d, of Δ^n (see Section 2.1 of [5] for a proof of this fact). Hence, if we subdivide each of them again, we will get simplices of diameter at most $dn^2/(n+1)^2$. Iterating this subdivision process k times, we get simplices of diameter at most $dn^k/(n+1)^k$. Since $n/(n+1)$ is less than 1, this sequence $dn^k/(n+1)^k$ tends to 0 as $k \to \infty$. Hence we can get any arbitrarily small diameter if we take k to be large enough. This is exactly what we need for the domain splitting result.

It is harder to see how to describe what we are doing algebraically. However, the language of chains turns out to be perfect for this. Let's look at the case $n = 2$, so we have a chain $s \in C_2(X)$ on a space X. For simplicity, assume that s is actually a simplex, i.e., a continuous map $s : \Delta^2 \to X$. By restricting s to any one of the smaller six triangles in the subdivision of Δ^2 we obtain another map to X, whose domain is a triangle. For clarity, let's label the triangles of the subdivision as $t_1, \ldots, t_6$, and, for $1 \leq j \leq 6$, let $i_j : t_j \to \Delta^2$ be the inclusion map of the triangle t_j in Δ^2. So the restriction of s to the triangle t_3, for example, is $s \circ i_3$. Now, each triangle is homeomorphic to Δ^2, and it will be important to fix the homeomorphism, so let $h_j : \Delta^2 \to t_j$ be such a homeomorphism, for $j = 1, \ldots, 6$. The composite $s \circ i_3 \circ h_3$ is then a continuous map $\Delta^2 \to X$, i.e., a 2-simplex on X. Hence we have six new simplices, $s \circ i_1 \circ h_1, \ldots, s \circ i_6 \circ h_6$. We can combine these to form a 2-chain on X by taking a suitable linear combination. As with the boundary operator, it turns out that we should take an alternating sum

$$s \circ i_1 \circ h_1 - s \circ i_2 \circ h_2 + \cdots - s \circ i_6 \circ h_6.$$

As all of these terms involve s, we can write this expression as

$$s \circ (i_1 \circ h_1 - i_2 \circ h_2 + \cdots - i_6 \circ h_6).$$

Each term inside the brackets is a map $\Delta^2 \to \Delta^2$, so together they form a 2-chain on Δ^2. We would call this chain $sd_2 \in C_2(\Delta^2)$, so that the subdivision of the 2-simplex $s : \Delta^2 \to X$ can be written as $s \circ sd_2$, and the subdivision of any 2-chain, $c \in C_2(X)$, is $c \circ sd_2$. However, note that we need to choose the homeomorphisms $h_1, \ldots, h_6$ carefully so that composing with sd_2 doesn't change the homology class.

Before we do that, we quickly note that this construction generalizes to all n quite easily. We can split Δ^n up inductively as above, and this gives inclusion maps i_j. Each n-simplex in the subdivision is homeomorphic to Δ^n and, as in the case $n = 2$, we need to choose homeomorphisms h_j carefully to preserve homology classes. If we can do that, then $sd_n = \sum_j \pm i_j \circ h_j$, and the subdivision, $sd(c)$, of an n-chain $c \in C_n(X)$ is given by $sd(c) = c \circ sd_n$.

In order to explicitly define the homeomorphisms h_j between Δ^n and the n-simplices in its subdivision, we need an explicit description of these n-simplices. The construction shows that there are $(n + 1)!$ of them (proof by induction: there are $n!$ $(n - 1)$-simplices in the subdivision of each face, and $n + 1$ faces), and rather than labelling them with the numbers $1, \ldots, (n+1)!$, they are more naturally indexed by permutations, as we will now see. Let $v_i \in \Delta^n$ be the vertex $(0, \ldots, 0, 1, 0, \ldots, 0)$ where the 1 is in the ith place (counting from 0, so that $v_0 = (1, 0, \ldots, 0)$). Given a permutation τ of the numbers $0, \ldots, n$, and a number i in the range $0, \ldots, n$, define a point $w_i^\tau \in \Delta^n$ by

$$w_i^\tau = \frac{1}{i+1}(v_{\tau(0)} + v_{\tau(1)} + \cdots + v_{\tau(i)}).$$

Thus, for example, $w_0^\tau = v_{\tau(0)}$, which is a vertex of Δ^n, and $w_1^\tau = (v_{\tau(0)} + v_{\tau(1)})/2$ which is halfway between two vertices of Δ^n, so is a vertex in the subdivision of Δ^n.

In the case $n = 2$, if label the vertices and 2-simplices as follows

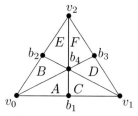

then the points w_i^τ, for the different permutations τ and indices i, are:

τ	w_0^τ	w_1^τ	w_2^τ	τ	w_0^τ	w_1^τ	w_2^τ
$(0,1,2)$	v_0	b_1	b_4	$(0,2,1)$	v_0	b_2	b_4
$(1,0,2)$	v_1	b_1	b_4	$(1,2,0)$	v_1	b_3	b_4
$(2,0,1)$	v_2	b_2	b_4	$(2,1,0)$	v_2	b_3	b_4

Note that for each permutation τ, the set of points w_0^τ, w_1^τ, w_2^τ are the vertices of one of the 2-simplices in the subdivision of Δ^2. For example, if $\tau = (0, 2, 1)$, then $w_0^\tau = v_0$, $w_1^\tau = b_2$, $w_2^\tau = b_4$ are the vertices of the simplex labelled B above. So every permutation gives one of these 2-simplices, and each 2-simplex is given by such a permutation.

All that remains is to define homeomorphisms between these 2-simplices of the subdivision, and Δ^2 itself. We do that by defining $h_\tau : \Delta^2 \to \Delta^2$ by $h_\tau(v_i) = w_i^\tau$ and interpolating between vertices. So, for example,

$$h_\tau\left(\frac{1}{3}v_1 + \frac{2}{3}v_2\right) = \frac{1}{3}w_1^\tau + \frac{2}{3}w_2^\tau.$$

This defines a map whose image is the 2-simplex corresponding to τ, which maps Δ^2 homeomorphically onto this 2-simplex. (So h_τ takes the place of the composite $i_j \circ h_j$ in our earlier discussion.) For example, $h_{(0,2,1)}$ can be depicted as

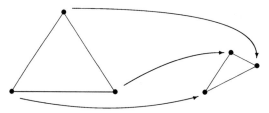

Since we are indexing these simplices by permutations, not by numbers, it is not so clear how to pick the signs in the alternating sum used to define sd_n. We do this using the **sign**, $\epsilon(\tau)$, **of the permutation** τ. This is given by the number of swaps required to turn τ into the identity; $\epsilon(\tau)$ is $+1$ if an even number of swaps are required, and -1 if an odd number of swaps are required (just like in Section 9.2). We then set

$$sd_n = \sum \epsilon(\tau) h_\tau \in C_n(\Delta^n),$$

where the summation runs over all permutations τ of $0, \ldots, n$.

The **barycentric subdivision** of a chain $s \in C_n(X)$, then, is the chain $\sigma_n(s) \in C_n(X)$ given by $\sigma_n(s) = s \circ sd_n$.

We need to check that this subdivision process does not change the homology class of the chain s. We will do this by constructing a chain homotopy from σ_n to the identity. To do that, we will need to relate the boundary of the subdivision chain sd_n to the subdivision of the boundary of Δ^n. In the case $n = 2$, we can calculate $\delta_2(sd_2) = \sum_i (-1)^i sd_2 \circ d^i$ explicitly and find that, after a lot of cancellation, we are left with

$$[v_0, b_1] - [v_0, b_2] - [v_1, b_1] + [v_1, b_3] + [v_2, b_2] - [v_2, b_3],$$

where $[v, w]$ is the simplex $\Delta^1 \to \Delta^2$ that sends $(x, y) \in \Delta^1$ to $xv + yw$. If $id_2 : \Delta^2 \to \Delta^2$ is the identity, then you can perform a similar calculation to find $\sigma_1(\delta_2(id_2))$, and you will see that it is exactly the same. This holds generally:

Lemma 10.13

For all $n \geq 0$, $\delta_n(sd_n) = \sigma_{n-1}(\delta_n(id_n))$, where $id_n : \Delta^n \to \Delta^n$ is the identity map.

Hence, for any space X, the family $\{\sigma_n : C_n(X) \to C_n(X)\}$ form a chain map.

Proof

To prove the first assertion, we note that $\delta_n(sd_n) = \sum_{i=0}^n (-1)^i sd_n \circ d^i$ and $sd_n \circ d^i = 0$ unless $i = n$, because the subsimplices in the subdivision which have the barycentre as a vertex pair up, with opposite signs, so they cancel. Then the remaining terms, $(-1)^n sd_n \circ d^n$, can be paired up with terms of

$$\sigma_{n-1}(\delta_n(id_n)) = \sigma_{n-1}\left(\sum_{i=0}^n (-1)^i d^i\right) = \sum_{i=0}^n (-1)^i \, d^i \circ sd_{n-1}$$

$$= \sum_{i=0}^n \sum_\tau (-1)^i \epsilon(\tau) d^i \circ h_\tau$$

in the following way. The term $h_\rho \circ d^n$ in $sd_n \circ d^n$ is equal to the term $d^i \circ h_\tau$ where $i = \rho(n)$ and τ is the permutation of $0, \ldots, n-1$ obtained from ρ by the rule

$$\tau(j) = \begin{cases} \rho(j) & \text{if } \rho(j) < \rho(n), \\ \rho(j) - 1 & \text{if } \rho(j) > \rho(n). \end{cases}$$

The second assertion is now a matter of plugging this identity (carefully) into the defining condition for a chain map: For any n-simplex $s \in C_n(X)$, we have:

$$\delta_n(\sigma_n(s)) = \delta_n(s \circ sd_n) = \sum_{i=0}^{n}(-1)^i (s \circ sd_n) \circ d_i = s \circ \left(\sum_{i=0}^{n}(-1)^i sd_n \circ d_i \right)$$
$$= s \circ \delta_n(sd_n),$$

and

$$\sigma_{n-1}(\delta_n(s)) = \sigma_{n-1}\left(\sum_{i=0}^{n}(-1)^i s \circ d_i \right) = \sum_{i=0}^{n}(-1)^i (s \circ d_i) \circ sd_{n-1}$$
$$= s \circ \left(\sum_{i=0}^{n}(-1)^i d_i \right) \circ sd_{n-1} = s \circ \delta_n(id_n) \circ sd_{n-1} = s \circ \sigma_{n-1}(\delta_n(id_n)).$$

By the first part, these two are equal. $\qquad\qquad\qquad\qquad\qquad\qquad\qquad\quad\square$

Now, to construct, for each space X, a chain homotopy from $\{\sigma_n : C_n(X) \to C_n(X)\}$ to the identity $\{\iota_n : C_n(X) \to C_n(X)\}$ means constructing a family of functions $\Phi_n^X : C_n(X) \to C_{n+1}(X)$ satisfying

$$\delta_{n+1}^X \circ \Phi_n^X + \Phi_{n-1}^X \circ \delta_n^X = \sigma_n - \iota_n.$$

Just as we did for σ_n and the boundary operator δ_n, we will do this by leaving X alone and working on the standard simplex Δ^n: We will construct a chain $F_n \in C_{n+1}(\Delta^n)$ and define Φ_n^X by $\Phi_n^X(s) = s \circ F_n$.

For each chain $s \in C_n(X)$, the defining property of Φ_n^X states that

$$\delta_{n+1}^X(\Phi_n^X(s)) + \Phi_{n-1}^X(\delta_n^X(s)) = \sigma_n(s) - \iota_n(s).$$

If $\Phi_n^X(s) = s \circ F_n$, then this equation becomes

$$\delta_{n+1}^X(s \circ F_n) + \delta_n^X(s) \circ F_{n-1} = \sigma_n(s) - s = s \circ (sd_n - id_n).$$

The left-hand side of this can be re-written as:

$$\delta_{n+1}^X(s \circ F_n) + \delta_n^X(s) \circ F_{n-1} = \sum_{i=0}^{n+1}(-1)^i(s \circ F_n) \circ d^i + \sum_{i=0}^{n}(-1)^i(s \circ d^i) \circ F_{n-1}$$
$$= s \circ \left(\sum_{i=0}^{n+1}(-1)^i F_n \circ d^i \right) + s \circ \left(\sum_{i=0}^{n}(-1)^i d^i \circ F_{n-1} \right)$$
$$= s \circ \left(\delta_{n+1}(F_n) + \delta_n(id_n) \circ F_{n-1} \right).$$

Hence we require the chains $\{F_n\}$ to satisfy

$$\delta_{n+1}(F_n) + \delta_n(id_n) \circ F_{n-1} = sd_n - id_n$$

for $n \geq 1$, and $\delta_1(F_0) = sd_0 - id_0$.

Since $sd_0 = id_0$ we can take F_0 to be 0. We then work inductively, using the same black-box approach[2] as in the proof of Theorem 10.11. If we have constructed $F_0, \ldots, F_{n-1}$ satisfying the above property, then

$$
\begin{aligned}
\delta_n(sd_n &- id_n - \delta_n(id_n) \circ F_{n-1}) \\
&= \delta_n(sd_n) - \delta_n(id_n) - \delta_n(\delta_n(id_n) \circ F_{n-1}) \\
&= \sigma_{n-1}(\delta_n(id_n)) - \delta_n(id_n) - \sum_{i=0}^{n}\sum_{j=0}^{n}(-1)^{i+j}d^j \circ F_{n-1} \circ d^i \\
&= \delta_n(id_n) \circ sd_{n-1} - \delta_n(id_n) - \sum_{j=0}^{n}(-1)^j d^j \circ \delta_n(F_{n-1}) \\
&= \delta_n(id_n) \circ (sd_{n-1} - id_{n-1}) - \delta_n(id_n) \circ \delta_n(F_{n-1}) \\
&= \delta_n(id_n) \circ (sd_{n-1} - id_{n-1} - (sd_{n-1} - id_{n-1} - \delta_{n-1}(id_{n-1}) \circ F_{n-2})) \\
&= \delta_n(id_n) \circ \delta_{n-1}(id_{n-1}) \circ F_{n-2} \\
&= (\delta_{n-1} \circ \delta_n)(id_n) \circ F_{n-2},
\end{aligned}
$$

which is 0 since $\delta_{n-1} \circ \delta_n = 0$.

Hence $sd_n - id_n - F_{n-1}\delta_n$ is in $\operatorname{Ker} \delta_n$ which, since Δ^n is acyclic, is the same as $\operatorname{Im} \delta_{n+1}$. Thus there is some element, which we call F_n, such that

$$\delta_{n+1}(F_n) = sd_n - id_n - \delta_n(id_n) \circ F_{n-1}.$$

This completes the inductive step and the construction of the chain homotopy F.

As usual, then, if $s \in \operatorname{Ker} \delta_n$, we have $\sigma_n(s) - s = F_{n-1}(\delta_n(s)) + \delta_{n+1}(F_n(s)) = \delta_{n+1}(F_n(s))$ which is, visibly, in the image of δ_{n+1}. Hence the difference will vanish when we pass to homology groups. In other words, s and $\sigma_n(s)$ give the same class in homology.

Thus we can prove the splitting result that we need:

Proposition 10.14 (Chain Splitting)

If $s \in C_n(X)$, and $X = U \cup V$ where U, V are open subsets of X, then there are elements $s_U \in C_n(U)$ and $s_V \in C_n(V)$ such that s is homologous to $s_U + s_V$. More precisely, if $i : U \to X$ and $j : V \to X$ are the inclusion maps, then there is some chain $t \in C_{n+1}(X)$ such that $s - C_n(i)(s_U) - C_n(j)(s_V) = \delta_{n+1}(t)$.

---

[2] This approach is called the **method of acyclic models**.

Proof

The chain s is, by definition, a linear combination of a finite number of singular simplices $f_1, \ldots, f_k : \Delta^n \to X$. The set Δ^n is compact, by the Heine–Borel theorem, being a closed, bounded subspace of $\mathbf{R}^{n+1}$. Therefore, we can use the domain splitting proposition, 6.29, for each simplex f_i and the cover U, V of X, to obtain a number $d_i > 0$ such that every subset of Δ^n of diameter less than d_i has image contained in either U or V. Since there are finitely many simplices $f_1, \ldots, f_k$, we can take $d = \min(d_1, \ldots, d_k)$ and this will still be a positive number. And when we apply any simplex f_i to any subset of X of diameter less than d, the image will be contained in either U or V.

Now we use the subdivision process, replacing s by $\sigma_n^i(s)$, where i is sufficient so that each simplex of $\sigma_n^i(s)$ is of diameter less than d. Then the image of each summand of $\sigma_n^i(s)$ will be contained either in U or in V. We group together those with image in U, and call this s_U, and group the remainder as s_V.   □

10.5 The Mayer–Vietoris Sequence

Proposition 10.14 provides the key to proving a homological analogue of the Van Kampen theorem. However, the flavour of the homological version is rather different, reflecting the more algebraic nature of homology. As before, we assume that we have a space X and two open subsets $U, V \subset X$ such that $X = U \cup V$. We then have inclusions $i : U \to X$ and $j : V \to X$ which induce homomorphisms $i_* : H_*(U) \to H_*(X)$ and $j_* : H_*(V) \to H_*(X)$. These can be combined to give a homomorphism $f : H_*(U) \oplus H_*(V) \to H_*(X)$ which we choose to define by

$$f(x, y) = i_*(x) - j_*(y).$$

The purpose of the negative sign will become clearer later.

As one might expect, we should also consider the intersection $U \cap V$ which has inclusion maps $k : U \cap V \to U$ and $l : U \cap V \to V$. Their induced homomorphisms can be combined into a map $g : H_*(U \cap V) \to H_*(U) \oplus H_*(V)$ defined by

$$g(x) = (k_*(x), l_*(x)).$$

Of course, the composites $i \circ k$ and $j \circ l$ from $U \cap V$ to X are identical, hence $i_* \circ k_* = j_* \circ l_*$. This shows that

$$f(g(x)) = f(k_*(x), l_*(x)) = i_*(k_*(x)) - j_*(l_*(x)) = 0,$$

i.e., Im $g \subset$ Ker f. In fact, Proposition 10.14 can be used to show that these two objects are equal:

Proposition 10.15

Im $g = $ Ker f.

Proof

We have already seen that Im $g \subset$ Ker f, so now we will prove the opposite inclusion, for which we suppose that $(c, d) \in H_n(U) \oplus H_n(V)$ is in Ker f. Let $\alpha \in C_n(U)$ and $\beta \in C_n(V)$ be cycles which represent c and d, respectively. If $f(c, d) = 0$, then $i_*(c) - j_*(d)$ is in Im δ_{n+1}, i.e. $C_n(i)(\alpha) - C_n(j)(\beta) = \delta_{n+1}(t)$ for some $t \in C_{n+1}(X)$. By Proposition 10.14, there are chains $t_U \in C_{n+1}(U)$, $t_V \in C_{n+1}(V)$ such that t is homologous to $C_{n+1}(i)(t_U) + C_{n+1}(j)(t_V)$, so that

$$\delta_{n+1}(t) = \delta_{n+1}(C_{n+1}(i)(t_U) + C_{n+1}(j)(t_V))$$
$$= C_n(i)(\delta_{n+1}(t_U)) + C_n(j)(\delta_{n+1}(t_V)).$$

Hence

$$C_n(i)(\delta_{n+1}(t_U)) + C_n(j)(\delta_{n+1}(t_V)) = C_n(i)(\alpha) - C_n(j)(\beta),$$

i.e.,

$$C_n(i)(\alpha - \delta_{n+1}(t_U)) = C_n(j)(\beta + \delta_{n+1}(t_V)).$$

The image of $C_n(i)$ in $C_n(X)$ consists of all n-chains whose images are contained inside U and, similarly the image of $C_n(j)$ consists of all chains whose images are contained inside V. The intersection of Im $C_n(i)$ and Im $C_n(j)$, therefore, consists of chains whose images are contained inside $U \cap V$. Hence both $\alpha - \delta_{n+1}(t_U)$ and $\beta + \delta_{n+1}(t_V)$ must have images in $U \cap V$, i.e.,

$$\alpha - \delta_{n+1}(t_U) = C_n(k)(x) \quad \text{and} \quad \beta + \delta_{n+1}(t_V) = C_n(l)(x)$$

for some n-chain $x \in C_n(U \cap V)$. Hence $c = [\alpha] = [\alpha - \delta_{n+1}(t_U)] = [C_n(k)(x)] = k_*(x)$, and $d = l_*(x)$ similarly. In other words, $(c, d) = g(x)$. $\qquad\square$

As in this proof the inclusion maps i and j, and their induced chain maps $C_n(i)$ and $C_n(j)$ can lead to rather clumsy expressions. Because of this, such maps are often omitted from formulas. So, for example, we may talk of a chain $t_U \in C_n(U)$ and then refer to t_U as a chain in X, meaning the chain $C_n(i)(t_U)$. For the remainder of this section we will use this convention, only mentioning the inclusions when this is necessary to avoid ambiguity.

We say that the sequence

$$H_*(U \cap V) \xrightarrow{\ g\ } H_*(U) \oplus H_*(V) \xrightarrow{\ f\ } H_*(X)$$

is **exact** at $H_*(U) \oplus H_*(V)$ to indicate that $\mathrm{Ker}\, f = \mathrm{Im}\, g$. If a longer sequence of homomorphisms is said to be exact, then this means that it is exact at each intermediate stage.

As well as using Proposition 10.14 to obtain information about the kernel of f as in Proposition 10.15, we can also use it to understand the image of f. In the Van Kampen theorem, 8.24, we needed $U \cap V$ to be path connected, i.e., $\pi_0(U \cap V) = 0$. In other words, to calculate $\pi_1(U \cup V)$, we needed to consider $\pi_0(U \cap V)$. The same sort of drop in degree occurs in the homological situation: To describe the image of f in $H_n(X)$, we have to look at $H_{n-1}(U \cap V)$.

Suppose we have a chain $c \in C_n(X)$ representing a class in $H_n(X)$. By Proposition 10.14, c is homologous to a sum $c_U + c_V$, with $c_U \in C_n(U)$, $c_V \in C_n(V)$. In particular, $\delta_n(c_U + c_V) = \delta_n(c) = 0$, i.e., $\delta_n(c_U) + \delta_n(c_V) = 0$. Hence $\delta_n(c_U) = -\delta_n(c_V)$. The left-hand side of this is a chain in U while the right-hand side is a chain in V. As they are equal, they must both be a chain in $U \cap V$. Hence, to every element $c \in C_n(X)$, we can assign an $(n-1)$-chain $\delta_n(c_U) \in C_{n-1}(U \cap V)$. This chain $\delta_n(c_U)$ is in $\mathrm{Ker}\, \delta_{n-1}$ since $\delta_{n-1}(\delta_n(c_U)) = 0$ and so determines an element of $H_{n-1}(U \cap V)$. (Note: Although $\delta_n(c_U)$ looks like it is in the image of δ_n, it is the image of a chain in U, not a chain in $U \cap V$. So, when we consider $\delta_n(c_U)$ as a chain in $U \cap V$, it need not be in $B_{n-1}(U \cap V)$.)

We would like to consider this construction as giving a function from $Z_n(X)$ to $H_{n-1}(U \cap V)$, but since the construction involved choosing elements c_U, c_V, we need to check that different choices do not lead to different answers.

Lemma 10.16

The class of $\delta_n(c_U)$ in $H_{n-1}(U \cap V)$ is dependent only on c, not on the choice of c_U.

Proof

Suppose we have c_U, c_V and c'_U, c'_V such that c is homologous to $c_U + c_V$ and $c'_U + c'_V$. So
$$c_U + c_V - (c'_U + c'_V) = \delta_{n+1}(d)$$
for some $d \in C_{n+1}(X)$. Using the chain splitting proposition on d, we find that, after subdividing enough times, we can find $d_U \in C_{n+1}(U)$ and $d_V \in C_{n+1}(V)$ such that d is homologous to $d_U + d_V$. In particular, $\delta_{n+1}(d) = \delta_{n+1}(d_U) + \delta_{n+1}(d_V)$, so
$$c_U - c'_U - \delta_{n+1}(d_U) = c'_V - c_V + \delta_{n+1}(d_V).$$
The left-hand side is a chain in U and the right-hand side is a chain in V, hence

they must both be a chain in $U \cap V$, with boundary

$$\delta_n(c_U - c'_U - \delta_{n+1}(d_U)) = \delta_n(c_U) - \delta_n(c'_U).$$

Hence the difference $\delta_n(c_U) - \delta_n(c'_U)$ is the image under δ_n of a chain in $U \cap V$. Thus the class of $\delta_n(c_U)$ in $H_{n-1}(U \cap V)$ does not depend on the choice of c_U. $\qquad \square$

So we can be confident that if you apply this construction to an element of $Z_n(X)$ and I apply it to the same element, then we will get the same class in $H_{n-1}(U \cap V)$.

The lemma also shows that if $c \in Z_n(X)$ is a boundary, i.e., $c \in B_n(X)$, then the construction gives the zero element in $H_{n-1}(U \cap V)$. Hence we can actually think of the construction as a function $\beta : H_n(X) \to H_{n-1}(U \cap V)$.

In fact, β is a homomorphism. For if $c, d \in H_n(X)$ can be split as $c \sim c_U + c_V$ and $d \sim d_U + d_V$, then we can split $c + d$ as $(c_U + d_U) + (c_V + d_V)$. The lemma then shows that this will give the same result for $\beta(c+d)$ as any other method. In other words, $\beta(c + d) = \delta_n(c_U + d_U) = \delta_n(c_U) + \delta_n(d_U) = \beta(c) + \beta(d)$.

Now, notice what happens when we compose β with f:

$$\beta(f(x,y)) = \beta(i_*(x) - j_*(y)),$$

where $x \in C_n(U)$, $y \in C_n(V)$. To evaluate $\beta(i_*(x) - j_*(y))$, we need to split $i_*(x) - j_*(y)$ as a sum of a chain in U and a chain in V. Obviously we can do that by taking x as our chain in U and $-y$ as our chain in V. Then $\beta(f(x,y))$ is the class of $\delta_n(x)$, and $\delta_n(x)$ is zero if x is a class in homology. Hence Im $f \subset$ Ker β. Again, these turn out to be equal.

Lemma 10.17

Im $f =$ Ker β.

Proof

We have already seen that Im $f \subset$ Ker β, so now we will prove the opposite inclusion, for which we suppose that $c \in C_n(X)$ is such that $\beta(c) = 0$. This means there is an element $d \in C_n(U \cap V)$ such that when we decompose c as $c \sim c_U + c_V$ with $c_U \in C_n(U)$ and $c_V \in C_n(V)$, we have $\delta_n(c_U) = \delta_n(d) = -\delta_n(c_V)$. Hence c is homologous to $c_U + c_V = (c_U - d) + (d + c_V)$, with $c_U - d \in C_n(U)$ and $d + c_V \in C_n(V)$ both being in Ker δ_n. Hence $c = f(c_U - d, -(d + c_V))$. $\qquad \square$

Reviewing what we have so far, we see that we have constructed maps f, g, β as follows:

$$H_n(U \cap V) \xrightarrow{g} H_n(U) \oplus H_n(V) \xrightarrow{f} H_n(X) \xrightarrow{\beta} H_{n-1}(U \cap V).$$

Notice that we have ended up in almost the same homology group as we started, just one degree down: H_{n-1} instead of H_n. For this reason, β is called the **connecting homomorphism**. And, just as this sequence is exact at $H_n(X)$ and exact at $H_n(U) \oplus H_n(V)$ so, if we carry it on, it will be exact at $H_{n-1}(U \cap V)$:

Lemma 10.18

Im $\beta = $ Ker g.

Proof

First note that

$$g(\beta(c)) = g(\delta_n(c_U)) = (k_* \delta_n(c_U), l_* \delta_n(c_U)) = (\delta_n(k_*(c_U)), \delta_n(l_*(c_U))$$

is in the image of δ_n, hence $g(\beta(c)) = 0$ in $H_n(U) \oplus H_n(V)$. So Im $\beta \subset$ Ker g.

Moreover, if $x \in H_{n-1}(U \cap V)$ is such that $g(x) = 0$, then $k_*(x) \in$ Im δ_n, and $l_*(x) \in$ Im δ_n, hence there are elements $c \in C_n(U)$, $d \in C_n(V)$ such that $k_*(x) = \delta_n(c)$, $l_*(x) = \delta_n(d)$. If this is the case, then, by construction, $\beta(c + d) = x$, i.e., $x \in$ Im β. □

Putting all of this together, we have

Theorem 10.19 (Mayer–Vietoris Theorem)

Given a space X and open subsets $U, V \subset X$ such that $X = U \cup V$, there is a long exact sequence

$$\cdots \to H_n(U \cap V) \to H_n(U) \oplus H_n(V) \to H_n(X) \to H_{n-1}(U \cap V) \to \cdots$$
$$\cdots \to H_1(X) \to H_0(U \cap V) \to H_0(U) \oplus H_0(V) \to H_0(X) \to 0$$

called the **Mayer–Vietoris sequence**.

This is very useful for calculating singular homology groups.

Example 10.20

The circle, S^1, can be decomposed as $S^1 = U \cup V$ where U is the complement of the North Pole, and V the complement of the South Pole. By stereographic projection, these are both homeomorphic to $\mathbf{R}$ and, hence, acyclic. Their intersection is a disjoint union of two spaces each, also, homeomorphic with $\mathbf{R}$. Using Corollary 10.6, the Mayer–Vietoris sequence becomes

$$\cdots \to 0 \to H_n(S^1) \to 0 \to \cdots \to 0 \to H_1(S^1) \to \mathbf{Z} \oplus \mathbf{Z} \to \mathbf{Z} \oplus \mathbf{Z} \to H_0(S^1) \to 0.$$

Since we know $H_0(S^1) = \mathbf{Z}$, as it has one path component, the kernel of this last map $\mathbf{Z} \oplus \mathbf{Z} \to H_0(S^1)$ must be $\mathbf{Z}$. By the same argument, the map $\mathbf{Z} \oplus \mathbf{Z} \to \mathbf{Z} \oplus \mathbf{Z}$ must have kernel $\mathbf{Z}$. And this must be equal to $H_1(S^1)$ since the kernel of the map $H_1(S^1) \to \mathbf{Z} \oplus \mathbf{Z}$ is zero. Hence $H_1(S^1) = H_0(S^1) = \mathbf{Z}$. For $n > 1$, $H_n(S^1)$ is sandwiched between two trivial groups and hence $H_n(S^1) = 0$ for all $n > 1$.

In this case, $H_n(S^1) = \pi_n(S^1)$ for all $n > 0$. Although homology can be thought of as an approximation to homotopy, it is very rarely this close.

Example 10.21

Similarly, we can express the 2-sphere as a union $S^2 = U \cup V$, where U is the complement of the North Pole, and V the complement of the South Pole. Both U and V are homeomorphic with $\mathbf{R}^2$, and their intersection is homeomorphic with $\mathbf{R} \times S^1$, which is homotopy equivalent to S^1. Hence $H_n(U \cap V) = 0$ if $n > 1$, and is equal to $\mathbf{Z}$ if $n = 0$ or 1. The Mayer–Vietoris sequence then gives

$$\cdots \to 0 \to H_n(S^2) \to 0 \to \cdots$$
$$\cdots \to 0 \to H_2(S^2) \to \mathbf{Z} \to 0 \oplus 0 \to H_1(S^2) \to \mathbf{Z} \to \mathbf{Z} \oplus \mathbf{Z} \to \mathbf{Z} \to 0.$$

Thus $H_n(S^2) = 0$ if $n > 2$ and $H_2(S^2) = \mathbf{Z}$. Finally, $H_1(S^2)$ must be zero, since the map $\mathbf{Z} \to \mathbf{Z} \oplus \mathbf{Z}$ has image $\mathbf{Z}$ and, so, must be injective.

In the same way, one can decompose any sphere S^n as a union of two open sets each homomeomorphic with $\mathbf{R}^n$, whose intersection is homotopy equivalent to S^{n-1}. Then, by induction on n, one sees that

$$H_i(S^n) = \begin{cases} \mathbf{Z} & \text{if } i = 0 \text{ or } i = n, \\ 0 & \text{otherwise.} \end{cases}$$

Note that $H_3(S^2) = 0$, in contrast with $\pi_3(S^2)$ which is non-zero by Example 8.8. The homology groups of spheres are significantly simpler than the homotopy groups of spheres, and so it is often easier to use them to prove other theorems, such as the following generalization of Brouwer's fixed-point theorem, 6.36.

Proposition 10.22 (Fixed-Point Theorem for the n-Disc)

If $f : D^n \to D^n$ is a continuous map, then it has a fixed point.

Proof

Using the same construction as in the proof for the case $n = 2$, we can turn a map $f : D^n \to D^n$ without fixed points into a map $g : D^n \to S^{n-1}$ which is the identity on the boundary. In other words, $g \circ i$ is the identity on S^{n-1}, where $i : S^{n-1} \to D^n$ is the inclusion map. On homology, $g \circ i$ then induces the identity. However, since $(g \circ i)_*$ is the same as $g_* \circ i_*$, this gives a factorization of the identity map on $H_m(S^{n-1})$ through $H_m(D^n)$:

$$H_m(S^{n-1}) \xrightarrow{i_*} H_m(D^n) \xrightarrow{g_*} H_m(S^{n-1}),$$

for any $m \geq 0$. Taking $m = n - 1$, we have $H_m(S^{n-1}) = \mathbf{Z}$, while $H_m(D^n) = 0$ by Example 10.7. So our factorization gives a composite

$$\mathbf{Z} \longrightarrow 0 \longrightarrow \mathbf{Z}$$

which is the identity on $\mathbf{Z}$. This cannot happen, hence the map f must have had a fixed point. $\qquad\square$

One consequence of this result concerns systems of equations. Suppose that we have a system of equations

$$f_1(x_1, \ldots, x_n) = 0,$$
$$f_2(x_1, \ldots, x_n) = 0,$$
$$\cdots$$
$$f_n(x_1, \ldots, x_n) = 0$$

of n equations in n variables, where the functions f_i are continuous on some subset of $\mathbf{R}^n$. A solution of this system corresponds to a fixed point of the function $h : \mathbf{R}^n \to \mathbf{R}^n$ whose ith coordinate is given by

$$h_i(x_1, \ldots, x_n) = x_i + f_i(x_1, \ldots, x_n).$$

If there is some closed disc D in $\mathbf{R}^n$ such that h is defined on D and $h(x) \in D$ for all $x \in D$, then the proposition tells us that h has a fixed point in D and, hence, that the system of equations has a solution in D.

Example 10.23

Let X be the figure of eight of Example 8.18. Let U be the subset obtained by deleting the leftmost point, and V the subset obtained by deleting the rightmost point:

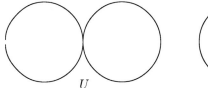 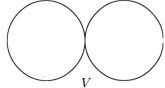

Then U and V are both homotopy equivalent to S^1, while their intersection is a cross, which is homotopy equivalent to a single point. Hence the Mayer–Vietoris sequence for X is

$$\cdots \to H_n(S^1) \oplus H_n(S^1) \to H_n(X) \to H_{n-1}(*) \to \cdots \to H_1(*) \to$$
$$H_1(S^1) \oplus H_1(S^1) \to H_1(X) \to H_0(*) \to H_0(S^1) \oplus H_0(S^1) \to H_0(X) \to 0.$$

Using our knowledge of the homology of S^1, this becomes

$$\cdots \to 0 \to H_n(X) \to 0 \to \cdots \to 0 \to \mathbf{Z} \oplus \mathbf{Z} \to H_1(X) \to \mathbf{Z} \to \mathbf{Z} \oplus \mathbf{Z} \to \mathbf{Z} \to 0.$$

This shows that $H_n(X) = 0$ for $n > 1$, while $H_1(X) = \mathbf{Z} \oplus \mathbf{Z}$.

In some situations the Mayer–Vietoris sequence cannot determine the homology precisely, although it can give some information about it.

Example 10.24

Let $X = T^2$ be the torus, $U, V \subset X$ the subsets

$$U = T^2 - \{(x, y, z) : (x - 2)^2 + z^2 = 1, \ y = 0\},$$
$$V = T^2 - \{(x, y, z) : (x + 2)^2 + z^2 = 1, \ y = 0\}.$$

So U and V are each obtained by removing a (different) circle from the torus, as shown below.

As these pictures make clear, U and V are both open cylinders, so homotopy equivalent to S^1. Their intersection, $U \cap V$, is homeomorphic to a disjoint union of two open cylinders, so is homotopy equivalent to a disjoint union of two circles. Thus the Mayer–Vietoris sequence is

$$\cdots \to H_n(S^1) \oplus H_n(S^1) \to H_n(T^2) \to H_{n-1}(S^1 \amalg S^1) \to \cdots$$
$$\to H_2(S^1) \oplus H_2(S^1) \to H_2(T^2) \to H_1(S^1 \amalg S^1) \to H_1(S^1) \oplus H_1(S^1) \to$$
$$H_1(T^2) \to H_0(S^1 \amalg S^1) \to H_0(S^1) \oplus H_0(S^1) \to H_0(T^2) \to 0,$$

i.e.,

$$\cdots \to 0 \to H_n(T^2) \to 0 \to \cdots \to 0 \to H_2(T^2)$$
$$\to \mathbf{Z} \oplus \mathbf{Z} \to \mathbf{Z} \oplus \mathbf{Z} \to H_1(T^2) \to \mathbf{Z} \oplus \mathbf{Z} \to \mathbf{Z} \oplus \mathbf{Z} \to H_0(T^2) \to 0.$$

This shows clearly that $H_n(T^2) = 0$ if $n > 2$, and we also know that $H_0(T^2) = \mathbf{Z}$ since T^2 is path connected. This implies that the map $\mathbf{Z} \oplus \mathbf{Z} \to H_0(T^2)$ has image $\mathbf{Z}$ and, hence, kernel $\mathbf{Z}$, so the preceding map $\mathbf{Z} \oplus \mathbf{Z} \to \mathbf{Z} \oplus \mathbf{Z}$ also has image $\mathbf{Z}$ and kernel $\mathbf{Z}$. Hence $H_1(T^2)$ has $\mathbf{Z}$ as a quotient. However, it could be that $H_1(T^2) = \mathbf{Z}$ and $H_2(T^2) = 0$. Or, it could be that $H_1(T^2) = \mathbf{Z} \oplus \mathbf{Z}$ and $H_2(T^2) = \mathbf{Z}$. In fact, there are other alternatives, too, such as $H_2(T^2) = \mathbf{Z} \oplus \mathbf{Z}$ and $H_1(T^2) = \mathbf{Z} \oplus \mathbf{Z}/m \oplus \mathbf{Z}/n$ for some integers m, n. This Mayer–Vietoris sequence cannot tell us which is the correct answer.

However, using other results, such as Theorem 10.27 below, it can be seen that $H_1(T^2) = \mathbf{Z} \oplus \mathbf{Z}$ and $H_2(T^2) = \mathbf{Z}$, just as for simplicial homology (Example 9.15).

10.6 Homology and Homotopy Groups

We originally constructed homology as a "rough approximation" to homotopy. One way of comparing the homotopy and homology groups of a space is by a group homomorphism, $h : \pi_n(X) \to H_n(X)$ called the "Hurewicz homomorphism" which we will now construct.

An element of $\pi_n(X)$ is a homotopy class of maps $S^n \to X$. If we take one map $f : S^n \to X$ representing a given homotopy class, then it induces a homomorphism on homology $H_n(S^n) \to H_n(X)$. Now, we have seen that $H_n(S^n) = \mathbf{Z}$, so this has a particularly important element, namely 1. The image of 1 under this induced homomorphism is, thus, an element of $H_n(X)$. We call this element $h_n(f)$.

Recall that if two continuous maps are homotopic, then they induce the same homomorphism on homology. So if we take any map $g : S^n \to X$ homotopic to f, then they will lead to the same induced homomorphism, and hence $h_n(f) = h_n(g)$. Thus our construction actually gives a function

$$h_n : \pi_n(X) \longrightarrow H_n(X).$$

Proposition 10.25

This function h_n is a homomorphism.

Proof

Recall from the start of Chapter 8 that the addition on $\pi_1(X)$ arises from pinching the points $(1,0)$ and $(-1,0)$ in S^1 together to obtain a figure of eight space. At the start of Section 8.1, we saw how this generalizes to give an addition on $\pi_n(X)$ by pinching the equator of S^n together, to give a space homeomorphic to two copies of S^n joined at the base point. (This is an example of a "wedge product", which we will meet properly in Chapter 11.) In other words, there is a pinching map $p : S^n \to (S^n \amalg S^n)/\sim$, where the equivalence relation identifies the base point in one copy of S^n with the base point in the other copy of S^n. For $n = 1$, we calculated the homology of this figure of eight space in Example 10.23, using the Mayer–Vietoris sequence. We can do a similar calculation for any n, and we find that $H_n(S^n \amalg S^n/\sim) = \mathbf{Z} \oplus \mathbf{Z}$. Moreover, the induced homomorphism $p_* : H_n(S^n) \to H_n(S^n \amalg S^n/\sim)$ is the homomorphism $\mathbf{Z} \to \mathbf{Z} \oplus \mathbf{Z}$ that sends 1 to $(1,1)$. And if $f, g : S^n \to X$ are pointed maps, so give rise to a map $(S^n \amalg S^n)/\sim \to X$, then the induced homomorphism for this map $\mathbf{Z}\oplus\mathbf{Z} \to H_n(X)$ takes $(1,1)$ to $f_*(1)+g_*(1)$. Hence $h_n(f\#g) = f_*(1) + g_*(1) = h_n(f) + h_n(g)$, i.e., h_n is a homomorphism. □

The function h_n is called the **Hurewicz homomorphism** in honour of its discoverer who also proved the following properties of it:

Theorem 10.26 (Hurewicz Theorem)

If $n > 1$ and $\pi_i(X) = 0$ for $0 \le i < n$, then h_n is an isomorphism. If $n = 1$ and $\pi_0(X) = 0$, then h_n is surjective and its kernel Ker $h_n \subset \pi_1(X)$ is generated by elements of the form $\alpha + \beta - \alpha - \beta$.

The proof is too long for this book, but it can be found in many books on algebraic topology, for example, in Section 4.2 of [5], or Chapters IV and VII of [2].

10.7 Comparison of Singular and Simplicial Homology

We have now seen two completely different definitions of homology groups. The simplicial theory is easy to calculate with, whereas in the singular theory it is quite hard to calculate the homology of many spaces. On the other hand, in the singular theory we have been able to prove theorems such as the existence of induced homomorphisms, their homotopy invariance, and the connection between homology and homotopy groups. These different advantages justify the inclusion of both theories, but is it reasonable that they should both be called "homology"?

In fact, many of the theorems that we have proved for the singular theory can also be proved for the simplicial theory, albeit with more effort. Continuous maps again have induced homomorphisms which depend only on the homotopy class of the map, and there is even an exact sequence akin to the Mayer–Vietoris sequence. Pursuing these similarities, Eilenberg and Steenrod enumerated six properties ("axioms") of homology theories and showed that both simplicial and singular homology have these properties. They also proved that, on triangulable spaces, every "homology" theory having these properties would give the same answer. In particular:

Theorem 10.27 (Eilenberg–Steenrod)

If X is a topological space homeomorphic to a simplicial complex K, then for each $i \geq 0$, the ith singular homology group $H_i(X)$ is isomorphic to the ith simplicial homology group $H_i(K)$.

After 50 years, their book [3] is still one of the best sources for the proof of this theorem.

The Eilenberg–Steenrod theorem says that, for triangulable spaces, the homology groups $H_i(X)$ depend only on the space X, not on the particular choice of homology theory. In particular, the simplicial homology of such a space is exactly the same as the singular homology of the space. This work had a tremendous unifying effect on algebraic topology and enables topologists to choose which homology theory they wish to work with to a large extent (and there are many more theories besides the two that we have met in this book). If there is a theorem you can prove in singular homology, then you can rely on its also being true for the simplicial homology of any nice space. And if there is a calculation you can perform in simplicial homology then, unless your space is particularly nasty, the result will also hold for singular homology. For example, we have proved that singular homology respects homotopy equivalences, therefore we can now deduce Theorem 7.13

Theorem 10.28 (Theorem 7.13)

If two triangulable spaces are homotopy equivalent, then they have the same Euler number.

Proof

If K and L are two triangulable spaces, then their simplicial homology agrees with their singular homology by the Eilenberg–Steenrod theorem. Since K and L are homotopy equivalent, their singular homology groups are isomorphic, by Corollary 10.10. Consequently, K and L have the same *simplicial* homology groups and so, by Proposition 9.9, they have the same Euler number.   □

EXERCISES

10.1. Using the Mayer–Vietoris sequence, or otherwise, calculate the singular homology of $\mathbf{R}^2 - \{0\}$ and of $\mathbf{R}^3 - \{0\}$.

10.2. Calculate the homology of $\mathbf{R}^2 - S^0 = \mathbf{R}^2 - \{(-1,0),(1,0)\}$. How does the homology change if you delete n distinct points from $\mathbf{R}^2$? What would be the homology if you deleted n points from $\mathbf{R}^m$?

10.3. Calculate the homology of $\mathbf{R}^2 - S^1$ using the Mayer–Vietoris sequence. What other space (with two components) has the same homology? Is this other space homotopy equivalent to $\mathbf{R}^2 - S^1$?

10.4. Calculate the homology of $\mathbf{R}\mathrm{P}^1$. (Hint: Example 5.6.)

10.5. Let V be the space obtained from the torus T^2 by deleting a small closed disc. Using the fact that this space is homotopy equivalent to a figure of eight, try to calculate the homology of T^2 using the Mayer–Vietoris sequence with this subspace V and an open disc U slightly larger than the hole in V.

10.6. Use the Mayer–Vietoris sequence to obtain information about the homology of the surface of genus two of Example 3.27. (Hint: The subspaces U and V should both be homeomorphic with the space V of Exercise 10.5.)

10.7. Calculate $H_*(\mathbf{Q})$.

11
More Deconstructionism

In Chapter 5, we studied the union, product and quotient constructions. These are elementary constructions, in the sense that they involve constructing a topological space directly. In this chapter, we will look at some non-elementary constructions which are built out of these elementary ones.

11.1 Wedge Products

Recall from Theorem 5.31 that two maps $f : X \to Z$, $g : Y \to Z$, with the same range space, can be combined to give a map $f \amalg g : X \amalg Y \to Z$ with the same range space again. If we work with *pointed* spaces and pointed maps, such as we did in Chapter 8, then this construction does not work so well, as $f \amalg g$ is not naturally a pointed map, not least because the disjoint union $X \amalg Y$ does not have an obvious base point. If X and Y are both pointed, then $X \amalg Y$ has *two* base points, and there is no way to pick one or the other of them.

To get around this, we take the quotient space where we identify these two base points together. Thus, if (X, x_0) and (Y, y_0) are two pointed spaces, then their the **wedge product**, or **one-point union**, $X \vee Y$ is the space defined by

$$X \vee Y = (X \amalg Y)/\!\sim$$

where two distinct points of $X \amalg Y$ are equivalent if, and only if, both are base points of either X or Y.

M.D. Crossley, *Essential Topology*, Springer Undergraduate
Mathematics Series, DOI 10.1007/978-1-84628-194-5_11,
© Springer-Verlag London Limited 2010

This has natural inclusions $i_X : X \to X \vee Y$ and $i_Y : Y \to X \vee Y$ which are both pointed maps, and such that any map $f : X \vee Y \to Z$ gives a pair of maps $f \circ i_X : X \to Z$, $f \circ i_Y : Y \to Z$. Moreover, any pair of pointed maps $f_X : X \to Z$, $f_Y : Y \to Z$ gives an unpointed map $f_X \amalg f_Y : X \amalg Y \to Z$ which preserves the equivalence relation $\sim$, i.e., $(f_X \amalg f_Y)(x) = (f_X \amalg f_Y)(x')$ if $x \sim x'$, hence, by Theorem 5.62, there is a corresponding map $X \vee Y \to Z$ which we denote by $f_X \vee f_Y$.

Example 11.1

If $X = Y = [0, 1]$, with 0 as the base point, then $X \vee Y$ is homeomorphic with a closed interval $[-1, 1]$ whose base point is in the middle, at 0. Depicting this wedge product as

explains where the $\vee$ symbol comes from.

Example 11.2

The figure of eight space of Example 8.18 is homeomorphic with $S^1 \vee S^1$.

Example 11.3

Similarly, we can take $S^n \vee S^n$ and form a "dumbbell" space. There is a continuous map $p : S^n \to S^n \vee S^n$ formed by pinching the equator of S^n to the base point, and hence a pair of pointed maps $f, g : S^n \to X$ gives a pointed map $(f \vee g) \circ p : S^n \to X$. As we saw in the proof of Theorem 10.26, this is another description of the map $f \# g$ of Chapter 8.

Example 11.4

The space $\mathbf{R}/\mathbf{Z}$ of Example 5.52 is a wedge product of countably many circles.

If X and Y are pointed simplicial complexes, then we can assume that the base point is a 0-simplex. For if it is not, then we can modify the simplicial complex by placing a vertex at the base point, and altering the other simplices as follows. The base point must be in the interior of one simplex, say an n-simplex, and we replace this n-simplex by $n + 1$ n-simplices, each given by omitting one vertex from the original n-simplex and replacing it by the base point. We do the same for all simplices that have this n-simplex as a subsimplex. For example, for $n = 2$, we replace one triangle (2-simplex) by three triangles:

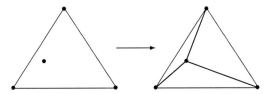

It is left as an exercise for the reader to verify that this does not change the homology (Exercise 11.1).

The wedge product $X \vee Y$ can then be given a simplicial structure in an obvious way, with n-simplices given by taking all the n-simplices of X and all the n-simplices of Y, i.e., $S_n(X \vee Y) = S_n(X) \cup S_n(Y)$, if $n > 0$, and the 0-simplices similarly, except that the base point of X must be identified with the base point of Y, i.e., $S_0(X \vee Y) = S_0(X) \cup S_0(Y)/x_0 \sim y_0$. In forming the simplicial homology of $X \vee Y$, the simplices of X and Y clearly do not interact apart from the 0-simplex common to both. This leads to the following calculation of the homology of their wedge product.

Theorem 11.5

If X and Y are two pointed, triangulable spaces, then $H_n(X \vee Y) = H_n(X) \oplus H_n(Y)$ for $n > 0$, and $H_0(X \vee Y) = H_0(X) \oplus H_0(Y)/\mathbf{Z}$. (Recall from Proposition 9.16 that the 0-th homology group is always a free abelian group).

11.2 Suspensions and Loop Spaces

If X is any topological space, then it is often useful to form the **cone** CX on X, namely the quotient $CX = X \times [0,1]/X \times \{0\}$.

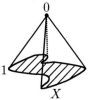

This is contractible, since the map $F : CX \times [0,1] \to CX$ defined by $F((x,s),t) = (x, st)$ is a homotopy from the identity map on CX to the constant map to the base point $X \times \{0\}$. However, it also contains a copy of X inside, as the subset $X \times \{1\}$ is homeomorphic to X.

We can also form the **double cone** DX on X, defined to be the quotient of $X \times [0,1]$ by the relation which identifies the subset $X \times \{0\}$ together to one point, and the subset $X \times \{1\}$ together to a different point. This can be thought of as two copies of CX joined along the copy of X contained in each.

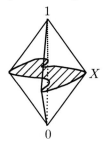

Example 11.6

If X is the circle S^1, then DX is homeomorphic with the sphere S^2.

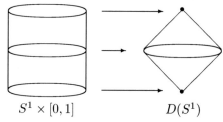

$$S^1 \times [0,1] \qquad\qquad D(S^1)$$

Similarly, $D(S^n)$ is homeomorphic to S^{n+1} for all non-negative n, even $n = 0$.

The homology groups of DX are very closely related to those of X:

Proposition 11.7

If $n > 1$, then $H_n(DX) = H_{n-1}(X)$, while $H_1(DX) = H_0(X)/\mathbf{Z}$ and $H_0(DX) = \mathbf{Z}$.

Proof

This is a simple calculation using the Mayer–Vietoris sequence. Let $U \subset DX$ be the image of $X \times [0, 2/3)$ under the quotient map $X \times [0,1] \to DX$, and let V be the image of $X \times (1/3, 1]$. Both of these are open subsets and both homotopy equivalent to the cone CX and, hence, contractible. The intersection $U \cap V$ is homeomorphic to $X \times (1/3, 2/3)$ and so homotopy equivalent to X.

Thus the Mayer–Vietoris sequence takes the form

$$\cdots \to 0 \oplus 0 \to H_n(DX) \to H_{n-1}(X) \to 0 \oplus 0 \to \cdots$$
$$\cdots \to 0 \oplus 0 \to H_1(DX) \to H_0(X) \to \mathbf{Z} \oplus \mathbf{Z} \to H_0(DX) \to 0.$$

Hence $H_n(DX) = H_{n-1}(X)$ for $n > 1$. This calculation also shows that $H_1(DX)$ is a subgroup of $H_0(X)$ with quotient $\mathbf{Z}$, i.e., $H_1(DX) = \mathbf{Z}^{r-1}$, where r is the number of path components in X, so $H_1(DX) = H_0(X)/\mathbf{Z}$. Finally, $H_0(DX) = \mathbf{Z}$ since DX is path connected.   □

Naturally, we would also like to know the homotopy groups of DX, but before we can do that we need to choose a base point. The space DX naturally comes with two base points, and there is no reason to choose one rather than the other. Moreover, if X is a pointed space, with base point x_0, then this provides another choice of base point for DX, thanks to the inclusion of X as $X \times \{1/2\} \subset DX$. To get around this choice, we work with the **suspension** of X, denoted by $\varSigma X$ and defined by

$$\varSigma X = \frac{X \times [0,1]}{X \times \{0,1\} \cup x_0 \times [0,1]} = \frac{DX}{\{x_0\} \times [0,1]}.$$

The name comes from thinking of X as being suspended between two points (0 and 1), much like DX. In fact, DX is sometimes called the **unreduced suspension**. The difference is that the dotted line, in the picture below, is collapsed to a single point in $\varSigma X$, but not in DX.

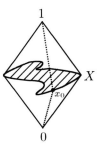

The following result relating DX and $\varSigma X$ is proved in Chapter VII, Section 1, of [2].

Proposition 11.8

If X is any triangulable space, then DX and $\varSigma X$ are homotopy equivalent.

Corollary 11.9

For $n \geq 0$, $\varSigma(S^n)$ is homotopy equivalent to S^{n+1}. In fact, they are homeomorphic, though this is easier to see from the pictures than to prove rigorously.

Corollary 11.10

If $n > 1$ and X is triangulable, then $H_n(\Sigma X) = H_{n-1}(X)$, while $H_1(\Sigma X) = H_0(X)/\mathbf{Z}$ and $H_0(\Sigma X) = \mathbf{Z}$.

The homotopy groups of a suspension ΣX cannot be described completely in terms of the homotopy groups of X. After all, this would inductively yield a description of the homotopy groups of spheres which, as we have seen, is a very difficult open problem. However, the following result, whose proof can be found in Section 4.2 of [5], describes some of the homotopy groups of ΣX:

Theorem 11.11 (Freudenthal Suspension Theorem)

If a triangulable space X is n-connected (i.e. $\pi_j(X) = 0$ for $j \leq n$), then $\pi_{i+1}(\Sigma X) = \pi_i(X)$ for $i \leq 2n$.

Example 11.12

If $X = S^2$, then X is 1-connected, so $\pi_{i+1}(\Sigma X) = \pi_i(X)$ if $i \leq 2$. In particular, $\pi_3(S^3) = \pi_3(\Sigma S^2) = \pi_2(S^2)$. Similarly, $\pi_n(S^n) = \pi_{n-1}(S^{n-1})$ for all $n \geq 3$ and hence, by induction, $\pi_n(S^n) = \pi_2(S^2)$ for all $n \geq 2$. In Example 11.24 below we will find that $\pi_2(S^2) = \pi_1(S^1) = \mathbf{Z}$, from which we can conclude that $\pi_n(S^n) = \mathbf{Z}$ for all $n > 0$.

The Freudenthal theorem shows that if X is n-connected, then $\pi_{i+1}(\Sigma X) = \pi_i(X)$ for $i \leq 2n$, so that, in particular, ΣX is $n + 1$-connected. Hence $\pi_{i+2}(\Sigma^2 X) = \pi_{i+1}(\Sigma X)$ for $i+1 \leq 2n+2$, i.e., a slightly larger range of i than before. Carrying this on indefinitely leads to the following result.

Corollary 11.13

If X is an n-connected, triangulable space, then $\pi_{i+j}(\Sigma^j X)$ is independent of j if $i \leq 2n + j$, i.e. $j \geq 2n - i$.

So if we take j large enough, then $\pi_{i+j}(\Sigma^j X) = \pi_{i+j+k}(\Sigma^{j+k} X)$ for all $k \geq 0$. Thus we define the **stable homotopy group** $\Pi_i(X)$ to be $\pi_{i+j}(\Sigma^j X)$ for j large enough for this group to be independent of the choice of j. This captures phenomena which are independent of dimension.

Closely related to the suspension construction is the "loop space" operator Ω. Given a pointed space Y, we define ΩY to be the set $\mathrm{Map}_*(S^1, Y)$ of pointed continuous maps $S^1 \to Y$. This needs to be topologized, and we do this in the following way. First, if $K \subset S^1$ is compact and $U \subset Y$ is open, then we consider the subset of ΩY consisting of all pointed, continuous maps $f : S^1 \to Y$ such that $f(K) \subset U$, and we define this subset to be open in the topology on ΩY. These open sets do not form a topology, so we also take all finite intersections of these open sets, and all arbitrary unions. The resulting topology on ΩY is called the **compact-open topology**, and ΩY is the **loop space** on Y. We make ΩY a pointed space by defining the base point to be the constant function $S^1 \to Y$ that takes every point in S^1 to the base point in Y.

The connection between Σ and Ω is as follows. Given a pointed map $f : X \to \Omega Y$, f assigns, to each point of X, a loop in Y, i.e., a map $[0,1] \to Y$. Hence, for each point $x \in X$ and $t \in [0,1]$, we get a point in Y, by evaluating the loop $f(x)$ at time t. We write $f(x,t)$ for this point. If $x = x_0$ is the base point, then $f(x)$ is the base point, i.e., the constant loop, so $f(x,t)$ is the base-point of Y, no matter what t is. Similarly, if $t = 0$, then $f(x,0)$ is the starting point of the loop $f(x)$, and this starting point is the base point of Y for all loops. Hence $f(x,0)$ is the base point of Y, no matter what x is. In other words, $f(x,t)$ defines a function $X \times [0,1] \to Y$ which sends $X \vee [0,1]$ to the base point. Hence we get a function $L(f) : \Sigma X \to Y$. Similarly, a function $g : \Sigma X \to Y$ gives a function $R(g) : X \to \Omega Y$, in such a way that $R(L(f)) = f$ and $L(R(g)) = g$, so these two operations are inverse to each other. Moreover, it can be seen that $L(f)$ is continuous if f is and that $R(g)$ is continuous if g is.

Hence, if we write $\mathrm{Map}_*(S,T)$ for the set of pointed, continuous maps from S to T, then we have a bijection

$$\mathrm{Map}_*(\Sigma X, Y) \longleftrightarrow \mathrm{Map}_*(X, \Omega Y).$$

We say that Σ and Ω are **adjoint functors**, meaning that if you apply one (Σ) to the domain, it is equivalent to applying the other (Ω) to the range.

This bijection preserves homotopies, i.e., $[\Sigma X, Z] = [X, \Omega Z]$ which shows that Ω interacts with homotopy as simply as Σ interacts with homology:

Theorem 11.14

If $n \geq 0$, then $\pi_n(\Omega Z) = \pi_{n+1}(Z)$.

On the other hand, the relation between Ω and homology is more complicated. In many cases it is possible to determine the homology of ΩX from that of X, but the process is too complicated for this book.

11.3 Fibre Bundles

We saw in Example 5.59 that the real projective space $\mathbf{RP}^n$ can be thought of as a quotient of the sphere S^n under the relation $\mathbf{x} \sim \mathbf{y}$ if $\mathbf{x} = \pm\mathbf{y}$. As with all quotient constructions, this means that we have a surjective map $\pi : S^n \to \mathbf{RP}^n$. In the case $n = 1$, we have seen how $\mathbf{RP}^1$ is homeomorphic with S^1, and this map π can be thought of as winding the circle twice around itself:

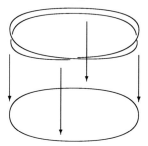

In general, as in the case $n = 1$, for every point $z \in \mathbf{RP}^n$, the preimage $\pi^{-1}(z)$ consists of two points, so these preimages are all essentially the same.

This might tempt us to think that we can rebuild S^n from $\mathbf{RP}^n$ by taking two points for each point in $\mathbf{RP}^n$. This sounds like the product space $S^0 \times \mathbf{RP}^n$ which has a copy of S^0 (i.e., two points) for each point in $\mathbf{RP}^n$.

In some limited sense, S^n does behave like this product space. But in other ways it is quite different. For example, S^n is connected whereas $S^0 \times \mathbf{RP}^n$ is not. And, whereas we can think of $S^0 \times \mathbf{RP}^n$ as having a copy of S^0 for each point in $\mathbf{RP}^n$, we can also think of it as having a copy of $\mathbf{RP}^n$ for each point in S^0. In other words, there is a symmetry between the two factors S^0 and $\mathbf{RP}^n$. It is this symmetry that is most strikingly lacking in the description of S^n as having two points for each point in $\mathbf{RP}^n$. If we think of having a copy of $\mathbf{RP}^n$ for each point in S^0, then it is clear that we have a disconnected space and, consequently, not S^n.

So we have a situation where a space S^n behaves somewhat like a product $(S^0 \times \mathbf{RP}^n)$ in that for each point in $\mathbf{RP}^n$ there is a copy of the other space S^0, but not if we swap these two spaces ($\mathbf{RP}^n$ and S^0) around. It turns out that this situation is very common and, while it is not as easy to work with as a product space, we can still learn a lot about the space from this sort of decomposition of it.

Another example is the Möbius band, M. There is a surjection π from M to the S^1, and the preimage $\pi^{-1}(\mathbf{x})$ of any point $\mathbf{x} \in S^1$ is homeomorphic to the closed interval $[0, 1]$. But the product $[0, 1] \times S^1$ is the cylinder. So we see that we have a "twisted" product because, where the cylinder can be thought of as a rectangle with two opposite edges glued together, in the Möbius band we have to put a twist in the rectangle before gluing the edges together. This is similar to the situation with $\mathbf{R}P^n$ and S^n described above, and in both cases we say we have a "fibre bundle":

Definition: A **fibre bundle** is a surjective, continuous map $p : E \to B$ together with a space F with the following property. For every point $x \in B$, there is an open set $U \subset B$ containing x, with a homeomorphism $\varphi_U : U \times F \to p^{-1}(U)$ such that the composite $p \circ \varphi_U : U \times F \to U$ is the projection $(u, f) \mapsto u$.

The space F is called the **fibre**, B is called the **base space** and E is called the **total space**. If $x \in B$, then $p^{-1}(x)$ is homeomorphic with F, via the restriction of φ_U^{-1} for the open set U containing x. In other words, E contains a copy of F for each point in B.

Example 11.15

The surjection $\pi : S^n \to \mathbf{R}P^n$ is a fibre bundle with fibre S^0. For any line $l \in \mathbf{R}P^n$, the preimage $\pi^{-1}(l)$ contains two points. If we take the open balls of radius 1 around each of these two points, and intersect with S^n, then we obtain an open subset of S^n which is the preimage under π of a certain subset of $\mathbf{R}P^n$. Since its preimage under π is open, this subset of $\mathbf{R}P^n$ is open, and we let it be U. Then $\pi^{-1}(U)$ is a disjoint union of two copies of U, i.e., $\pi^{-1}(U) \cong U \times S^0$, with the restriction of π to $\pi^{-1}(U)$ given by $(u, s) \mapsto u$, exactly as required.

Example 11.16

The surjection $\pi : M \to S^1$, from the Möbius band to the circle, is a fibre bundle with fibre $[0, 1]$.

Example 11.17

Any product space $S \times T$ gives rise to two fibre bundles: One $S \times T \to T$ with fibre S, and another $S \times T \to S$ with fibre T. These are called **trivial fibre bundles**.

Example 11.18

The surjection $e : \mathbf{R} \to S^1$ is a fibre bundle with fibre $\mathbf{Z}$.

Suppose now that B is a pointed space, with base point b_0. Then the fibre F is homeomorphic to the preimage $p^{-1}(b_0)$ of this point. Since $p^{-1}(b_0)$ is a subset of E, this gives us an inclusion $i : F \hookrightarrow E$. The composite $p \circ i : F \to B$

will, then, take every point of F to b_0. In other words, this composite is a constant map. If we assume, in addition, that E is pointed and that the map p preserves the base point, then the base point of E is in $p^{-1}(b_0)$. Thus we can give F a base point – the point corresponding to the base point in $p^{-1}(b_0)$, so that the whole construction can be carried out in the world of pointed spaces and pointed maps.

Thus, when we take homotopy groups, we get a sequence

$$\pi_n(F) \xrightarrow{i_*} \pi_n(E) \xrightarrow{p_*} \pi_n(B)$$

for each $n \geq 0$, where the composite $p_* \circ i_* = (p \circ i)_*$ is the zero homomorphism. In fact, it can be shown that this sequence is exact: If $p_*[f] = b_0$, then f is homotopic to a map whose image is contained in $p^{-1}(b_0) = F$, i.e., $[f]$ is in the image of i_*. Proofs of this, and the other exactness results of this section, are given in Section 4.2 of [5] and Chapter VII, Section 6 of [2].

Moreover we can join all these sequences together into one long exact sequence, via a map $\pi_n(B) \to \pi_{n-1}(F)$ for each $n > 0$. We have seen one instance of this already, for the fibre bundle $e : \mathbf{R} \to S^1$, which we used to calculate $\pi_1(S^1)$. There we saw that a map $f : D^1 \to S^1$ can be lifted to a map $\tilde{f} : D^1 \to \mathbf{R}$ such that $e \circ \tilde{f} = f$. This generalizes as follows.

Theorem 11.19 (Lifting Theorem)

Given a fibre bundle $p : E \to B$ and a map $f : D^n \to B$, there is a lift $\tilde{f} : D^n \to E$ such that $p \circ \tilde{f} = f$, i.e., there is a dotted map to make the following diagram commute.

Proof

If p is a fibre bundle, then B has a covering by open sets U such that $p^{-1}(U)$ is homeomorphic with $U \times F$ in such a way that p acts as the projection $U \times F \to U$. By the domain splitting proposition, 6.29, we can divide D^n into small enough n-cubes so that each maps into one of these subsets U. We then define $\tilde{f}$ piece by piece, starting from one corner, as in Proposition 6.28. At each stage we are faced with the problem of having a map $g : D^n \to U$ (g being some restriction of f) and a map $\tilde{g} : V \to U \times F$ defined on some subspace V of D^n, satisfying $g = p \circ \tilde{g}$, and we must extend $\tilde{g}$ over the whole of D^n. Since the range of $\tilde{g}$ is a product $U \times F$, we can consider $\tilde{g}$ as a pair of maps $\tilde{g}_1$, with range U, and $\tilde{g}_2$, with range F, by Theorem 5.42. Now $\tilde{g}_1$ is exactly $p \circ \tilde{g}$, so we must take $\tilde{g}_1$ to be g, and we have only to deal with $\tilde{g}_2$. (In the case of Proposition 6.28, F was $\mathbf{Z}$, so $\tilde{g}_2$ was constant and easily dealt with.) In fact, the subspace V can only take certain forms: It must contain the corner point $(0, \ldots, 0)$ of D^n and some combination of faces, edges etc. adjacent to this corner. The details are not very illuminating, but it is always possible to define a map from D^n to V which restricts to the identity on V; hence we can extend $\tilde{g}_2$ over the whole of D^n by composing $\tilde{g}_2|_V$ with such a map.   □

Now, if we are given a pointed map $S^n \to B$, then we can consider this as a map $f : D^n \to B$ that sends the boundary of D^n to the base point of B, and lift this to a map $\tilde{f} : D^n \to E$. This must map the boundary of D^n to $p^{-1}(B)$, which is homeomorphic with F. Hence we can consider this as a map $S^{n-1} \to F$. Moreover, if two maps $S^n \to B$ are homotopic, then the homotopy lifts to give a homotopy between the resulting maps $S^{n-1} \to F$. Hence we have a function $\gamma : \pi_n(B) \to \pi_{n-1}(F)$. Looking back at how addition on π_n is defined, it can be seen that γ is a group homomorphism.

Theorem 11.20

Given a fibre bundle $p : E \to B$, with fibre F, the sequence

$$\cdots \to \pi_n(F) \xrightarrow{i_*} \pi_n(E) \xrightarrow{\pi_*} \pi_n(B) \xrightarrow{\gamma} \pi_{n-1}(F) \to \cdots \to \pi_0(E) \to \pi_0(B)$$

is exact, meaning that the image of each function is the kernel of the following function. (The last four functions in this sequence are not group homomorphisms, because their range is not a group. However each range set has a "zero" element, namely the component of the base point, and we use the term "kernel" to denote the preimage of this zero element.) This is called the **long exact sequence of a fibre bundle** or **homotopy exact sequence of a fibre bundle**.

A proof of this theorem can be found in Section 4.2 of [5] or Chapter VII, Section 6, of [2].

Example 11.21

Using the fibre bundle $e : \mathbf{R} \to S^1$ with fibre $\mathbf{Z}$, we get the long exact sequence

$$\cdots \to \pi_n(\mathbf{Z}) \to \pi_n(\mathbf{R}) \to \pi_n(S^1) \to \pi_{n-1}(\mathbf{Z}) \to \pi_{n-1}(\mathbf{R}) \to \cdots$$
$$\cdots \to \pi_1(\mathbf{R}) \to \pi_1(S^1) \to \pi_0(\mathbf{Z}) \to \pi_0(\mathbf{R}) \to \pi_0(S^1).$$

Now, if $n > 0$, then $\pi_n(\mathbf{Z})$ consists of just one element, because the only pointed map $S^n \to \mathbf{Z}$ is the constant map to 0. And $\pi_n(\mathbf{R}) = 0$ if $n > 0$, because all maps $S^n \to \mathbf{R}$ are homotopic. On the other hand, $\pi_0(\mathbf{Z}) = \mathbf{Z}$, because there is one pointed map $S^0 \to \mathbf{Z}$ for each element of $\mathbf{Z}$ and none are homotopic. At the other extreme, $\pi_0(\mathbf{R})$ has a single element, because all pointed maps $S^0 \to \mathbf{R}$ are homotopic.

Hence the sequence actually looks like

$$\cdots \to 0 \to 0 \to \pi_n(S^1) \to 0 \to 0 \to \cdots \cdots \to 0 \to \pi_1(S^1) \to \mathbf{Z} \to 0 \to \pi_0(S^1),$$

from which we see that $\pi_1(S^1) = \mathbf{Z}$ and $\pi_n(S^1) = \{0\}$ if $n \neq 1$, as stated in Example 8.6. Of course, this sequence is merely a tidier way of presenting the calculation carried out earlier.

Example 11.22

The real projective space $\mathbf{RP}^m$ can be expressed as the quotient of S^m where opposite points are identified, since these both lie on the same line through the origin. Putting that another way, we have a map $p : S^m \to \mathbf{RP}^m$, which sends a point on the sphere to the line through the origin and that point. This is a fibre bundle, and the fibre over any point consists of just two elements, so is homeomorphic to S^0. Thus we have a long exact sequence

$$\cdots \to \pi_n(S^0) \to \pi_n(S^m) \to \pi_n(\mathbf{RP}^m) \to \pi_{n-1}(S^0) \to \pi_{n-1}(S^m) \to \cdots$$
$$\cdots \to \pi_1(S^m) \to \pi_1(\mathbf{RP}^m) \to \pi_0(S^0) \to \pi_0(S^m) \to \pi_0(\mathbf{RP}^m).$$

Since S^n is connected if $n > 0$, so $\pi_n(S^0) = 0$, and $\pi_0(S^0)$ consists of two elements. Thus, for $n > 1$, the sequence shows that $\pi_n(\mathbf{RP}^m) = \pi_n(S^m)$, which is, unfortunately, hard to calculate. For π_1 and π_0 we have a sequence

$$\cdots 0 \to \pi_1(S^m) \to \pi_1(\mathbf{RP}^m) \to \pi_0(S^0) \to 0 \to 0.$$

If $m > 1$, then $\pi_1(S^m) = 0$, and so $\pi_1(\mathbf{RP}^m)$ has two elements. The only group with two elements is $\mathbf{Z}/2$, so $\pi_1(\mathbf{RP}^m) = \mathbf{Z}/2$ if $m > 1$. (If $m = 1$, then $\mathbf{RP}^1 \cong S^1$ so $\pi_1(\mathbf{RP}^1) = \mathbf{Z}$.)

Corollary 11.23

Since we have seen (in Proposition 5.67) that $SO(3)$ is homeomorphic with $\mathbf{RP}^3$, this shows that $\pi_1(SO(3)) = \mathbf{Z}/2$.

In other words, there are paths in $SO(3)$ which are not null-homotopic but which, when traced around twice, become null-homotopic.

This can be envisaged in various ways. For example, if one person holds one end of a scarf, and another person twists the other end through $360°$, then it is not possible to untwist the scarf without moving either end. However, if it is twisted through $720°$, then the scarf can be untwisted while keeping boths ends fixed, by looping the middle of the scarf around one of the ends. Chapter III of [2] contains some more discussion of this, together with some helpful photos of a demonstration of this twisting phenomenon.

Example 11.24

In Example 8.8, we met the Hopf map $H : S^3 \to S^2$. Recall that this is defined as the map $(z_1, z_2) \mapsto z_1/z_2$, where $S^3 = \{(z_1, z_2) \in \mathbf{C}^2 : |z_1|^2 + |z_2|^2 = 1\}$, and $S^2 \cong \mathbf{C} \cup \{\infty\}$ as in Example 5.7. The preimage of any point z is then the set of elements $(z_1, z_2) \in S^3$ such that $z_1/z_2 = z$, i.e., $z_1 = z z_2$. With the restriction that $|z_1|^2 + |z_2|^2 = 1$, this gives a circle. Developing this a little further, we see that H is a fibre bundle with fibre S^1. Hence we get the long exact sequence

$$\cdots \to \pi_n(S^1) \to \pi_n(S^3) \to \pi_n(S^2) \to \pi_{n-1}(S^1) \to \pi_{n-1}(S^3) \to \cdots$$
$$\cdots \to \pi_1(S^3) \to \pi_1(S^2) \to \pi_0(S^1) \to \pi_0(S^3) \to \pi_0(S^2).$$

Using the fact that $\pi_i(S^j) = 0$ if $i < j$ (see Example 8.7), we have $\pi_2(S^3) = 0$ and $\pi_1(S^3) = 0$, hence $\pi_2(S^2) = \pi_1(S^1) = \mathbf{Z}$. Having established this, the Freudenthal suspension theorem 11.11 shows that $\pi_n(S^n) = \mathbf{Z}$ for all $n > 0$, as mentioned earlier, in Example 11.12. In particular, $\pi_3(S^3) = \mathbf{Z}$. Then, because $\pi_3(S^1) = \pi_2(S^1) = 0$, this sequence also shows that $\pi_3(S^3) = \pi_3(S^2) = \mathbf{Z}$, being generated by the image under H_* of the identity map. Since this image is exactly the class of H, we see that $\pi_3(S^2)$ is generated by the Hopf map $H : S^3 \to S^2$.

11.4 Vector Bundles

A particularly important type of fibre bundle is one where the preimage of each point in the base space has the structure of a vector space in a compatible way. This is called a "vector bundle." More precisely: A **vector bundle** is a fibre bundle $p : E \to B$ where the fibre F is a vector space, and each preimage $p^{-1}(b)$ is given the structure of a vector space in such a way that the homeomorphism $\phi_U : U \times F \to p^{-1}(U)$ restricts to a linear isomorphism $\{b\} \times F \to p^{-1}(b)$.

Example 11.25

If we take an infinite cylinder, $S^1 \times \mathbf{R}$, then the projection $p : S^1 \times \mathbf{R} \to S^1$ is a fibre bundle with fibre $\mathbf{R}$ which has an obvious vector space structure. For any subset $U \subset S^1$, $p^{-1}(U) = U \times \mathbf{R} \subset S^1 \times \mathbf{R}$ and we can take the homeomorphism $\phi_U : U \times \mathbf{R} \to p^{-1}(U)$ to be the identity map. Then the restriction of ϕ_U to $\{x\} \times \mathbf{R}$ is the identity on $\mathbf{R}$, which is certainly a linear isomorphism.

Example 11.26

We can make an infinite Möbius band into a vector bundle similarly: Take $[0,1] \times \mathbf{R}$, and join the ends of $[0,1]$ together with a twist. The projection $p : M \to S^1$ is a fibre bundle with fibre $\mathbf{R}$, and each preimage $p^{-1}(x)$ has a natural vector space structure that is compatible with that on the fibre $\mathbf{R}$ in the required way.

Example 11.27

Let TS^2 be the set of all tangent vectors on S^2, so a point in TS^2 is a point of S^2, together with a vector in $\mathbf{R}^3$, starting from that point of S^2, and parallel to the surface of S^2 at that point. For example, the set of tangent vectors around the North Pole in S^2 is in one-to-one correspondence with the subset $\{(x, y, 1) : x, y \in \mathbf{R}\}$ of $\mathbf{R}^3$. The projection $TS^2 \to S^2$ sends a tangent vector to its start point. Then for each point of S^2, the set of tangent vectors forms a space homeomorphic with $\mathbf{R}^2$ which can, thus, be given a real vector space structure. This is called the **tangent bundle** of S^2.

Example 11.28

Similarly, by taking all the normal vectors to S^2, we obtain the **normal bundle** $NS^2 \to S^2$. At each point in S^2, there is only one direction that is normal to the surface, so the fibre of this bundle is one-dimensional, i.e., the fibre is $\mathbf{R}$.

The set of different vector bundles over a given base space is a reflection of the topology of that base space. So one way of studying the base space is to consider all vector bundles over it. To make this more manageable, we will consider some vector bundles as equivalent.

If $\pi_1 : E_1 \to B$ and $\pi_2 : E_2 \to B$ are two vector bundles over the same space B, then they are **equivalent** if there is a homeomorphism $f : E_1 \to E_2$ such that $\pi_2 \circ f = \pi_1$. It is simple to check that this is an equivalence relation, so we can consider equivalence classes of vector bundles. Let $V(X)$ be the set of equivalence classes of real vector bundles over a given space X.

If we take two real vector spaces V, W, then we can form their sum $V \oplus W$, whose dimension will be $\dim V + \dim W$ if these are both finite. With some care, one can perform a similar operation on vector bundles. Thus, if we have two real vector bundles over the same base space B, we can form their direct sum, which will again be a vector bundle over B, and the fibre will be the direct sum of the fibres of the original vector bundles. This gives an addition operation on $V(X)$ since this addition respects the equivalence relation.

It is convenient to also be able to subtract, so as to have a group, and we can do this by allowing "virtual bundles." To see how to do this, consider how we can build the integers out of the natural numbers. We take the set $\mathbf{N}^2$ of pairs of natural numbers, where we want a pair (m, n) to represent $m - n$. So we need to identify certain pairs: We consider (m, n) and (p, q) to be equivalent if $m + q = p + n$. Thus, for example, any pair of the form (m, m) is equivalent to any other pair of this form, and these form the zero element of $\mathbf{Z}$. Addition is defined by $(m, n) + (p, q) = (m + p, n + q)$ and, as one can check, this preserves the identifications that we have imposed.

Thus, by considering equivalence classes of pairs of vector bundles over X, we can form an Abelian group, which is written $K(X)$, the K-**theory** of the space X. Many more details about $K(X)$ can be found in [1] and [6].

If we have a continuous map $f : X \to Y$, then a vector bundle over Y can be pulled back to give a vector bundle over X. To see this, let $p : E \to Y$ be a vector bundle, and define E_f to be the set

$$E_f = \{(x, e) \in X \times E : f(x) = p(e)\},$$

topologized as a subspace of the product space $X \times E$. The projection $p_f : E_f \to X$ takes an element (x, e) to $x \in X$. This is surjective because if $x \in X$, then $f(x) \in Y$ and since p is surjective, there is therefore some $e \in E$ such that $p(e) = y$. Hence $(x, e) \in E_f$ and $p_f(x, e) = x$.

$$
\begin{array}{ccc}
E_f & & E \\
\downarrow{\scriptstyle p_f} & & \downarrow{\scriptstyle p} \\
X & \xrightarrow{\ \ f\ \ } & Y
\end{array}
$$

Moreover, $p_f : E_f \to X$ is a fibre bundle, with fibre

$$p_f^{-1}(x) = \{(x, e) : p(e) = f(x)\} = \{x\} \times p^{-1}(f(x)).$$

This bundle p_f is the **pull-back** of the bundle $p : E \to Y$ along the map $f : X \to Y$.

The pull-back construction gives rise to an induced map $f_* : K(Y) \to K(X)$ which turns out to be a homomorphism of Abelian groups. Note that, unlike the induced homomorphism on homotopy groups or homology groups, this induced homomorphism goes in the opposite direction to the original continuous map. Consequently, we say that K-theory is **contravariant**, to indicate that it reverses the direction of maps.

Thus we have an operation which takes a topological space and gives a group, and takes a continuous map and gives a group homomorphism, much like the homotopy groups and homology groups do. In fact, $K(X)$ can be developed into something that behaves rather like homology and is an example of a "generalized cohomology theory." But that's the start of another book.

EXERCISES

11.1. Verify that subdividing an n-simplex to include a base point, as in Section 11.1, does not change the homology of the simplicial complex.

11.2. Use the results of this chapter to simplify as many of the calculations of homology and homotopy groups given in this book as possible.

Solutions to Selected Exercises

Chapter 2

2.1. For $x \in (2,3)$, let $\delta_x = \min(x-2, 3-x)$.

2.3. $\mathbf{R} - \{x\} = (-\infty, x) \cup (x, \infty)$. If $y \in (-\infty, x)$, then let $\delta_y = x - y$ so that $\delta_y > 0$ and $(y - \delta_y, y + \delta_y) \subset (-\infty, x)$. Hence $(-\infty, x)$ is open, and so is (x, ∞) similarly. Hence $\mathbf{R} - \{x\}$ is open by Proposition 2.4, so $\{x\}$ is closed.

2.5. $f^{-1}[-2,2] = [-2,2]$, $f^{-1}(2,18) = (2,3)$, $f^{-1}[2,18] = \{-1\} \cup [2,3)$, $f^{-1}[0,2] = [-\sqrt{3},0] \cup [\sqrt{3},2]$.

Chapter 3

3.1. (1) 4. (2) At least 19.

3.2. Yes. No. Yes. No.

3.4. If f is continuous and $C \subset T$ is closed, then $T - C$ is open so $f^{-1}(T-C)$ is open, and $f^{-1}(T-C) = S - f^{-1}(C)$ is open, so $f^{-1}(C)$ is closed. The converse is proved similarly.

3.8. It is not continuous, since $\{0\}$ is an open subset of $\mathbf{Z}$ but $f^{-1}(0) = [0,1)$ which is not an open subset of $\mathbf{R}$.

3.10. If $w^2 + x^2 + y^2 + z^2 = 1$, then $w^2 \leq 1$, so $-1 \leq w \leq 1$, and we can write $w = \cos\theta$. Then $x^2 + y^2 + z^2 = \sin^2\theta$, so if $x' = x/\sin\theta$ etc., then $(x')^2 + (y')^2 + (z')^2 = (x^2 + y^2 + z^2)/\sin^2\theta = 1$, i.e., $(x', y', z') \in S^2$.

M.D. Crossley, *Essential Topology*, Springer Undergraduate
Mathematics Series, DOI 10.1007/978-1-84628-194-5,
© Springer-Verlag London Limited 2010

Chapter 4

4.2. 1) Connected, not Hausdorff. 2) Disconnected, not Hausdorff. 3) Disconnected, Hausdorff.

4.3. Let $f : \mathbf{R} \to \mathbf{Q}$ be continuous, and let $S \subset \mathbf{Q}$ be the image of f. If f is not constant, then S contains at least two points s_1, s_2. Hence S is disconnected since between any two rationals there is an irrational number. Taking the subspace topology on S allows us to restrict f to a surjection $\mathbf{R} \to S$ from a connected to a disconnected space, and this cannot happen by Proposition 4.11.

4.6. det is a continuous surjection $\mathrm{GL}(3, \mathbf{R}) \to \mathbf{R} - \{0\}$ and $\mathbf{R} - \{0\}$ is not compact (since, for example, there is an unbounded function $\mathbf{R} - \{0\} \to \mathbf{R}$, namely $x \mapsto 1/x$). Hence $\mathrm{GL}(3, \mathbf{R})$ is not compact. Both $\mathrm{O}(3)$ and $\mathrm{SO}(3)$ are compact; this can be proved using the Heine–Borel theorem.

4.9. It would be connected, compact, but not Hausdorff.

Chapter 5

5.3. Let (a, b) be an interval in $\mathbf{R}$. Since tan is increasing on the range $(-\pi/2, \pi/2)$, if $\tan^{-1}(a) < x < \tan^{-1}(b)$ then $a < \tan(x) < b$. Hence $f^{-1}(a, b) = (2\tan^{-1}(a)/\pi, 2\tan^{-1}(b)/\pi)$, which is open. Thus f is continuous.

5.7. We can define homeomorphisms $f : C \to A$ by $f(x, y, z) = ((1 + z)x, (1 + z)y)$ and $g : A \to C$ by $g(x, y) = (\frac{x}{\sqrt{x^2+y^2}}, \frac{y}{\sqrt{x^2+y^2}}, \sqrt{x^2 + y^2} - 1)$.

5.9. Suppose that S and T are Hausdorff, and (s, t), (s', t') are two distinct points in $S \times T$. If $s \neq s'$, then there are two non-overlapping open subsets Q, Q' of S such that $s \in Q$ and $s' \in Q'$. The sets $Q \times T$ and $Q' \times T$ are then non-overlapping open subsets of $S \times T$, and $(s, t) \in Q \times T$ and $(s', t') \in Q' \times T$. If $s = s'$, then $t \neq t'$ and a similar argument can be used.

Suppose $S \times T$ is compact. To show S is compact, let $\mathcal{U}$ be any open covering of S. For each set $U \in \mathcal{U}$, the set $U \times T$ in $S \times T$ is open as both U and T are open. And every point in $S \times T$ will lie in one set $U \times T$; hence we have a cover of $S \times T$. As this space is compact, we can refine this cover, and the corresponding U spaces give a refinement of the original cover of S. Similarly, T is compact.

If $S \times T$ is connected, then S and T must be connected as the projections $S \times T \to S$ and $S \times T \to T$ are continuous surjections.

5.11. T^2 and G_2.

Chapter 6

6.2. There is only one homotopy class of maps $(0, 1) \to (0, 1)$.

6.4. If f is not surjective, then its image lies in $S^n - \{s\}$ for some point $s \in S^n$. By stereographic projection, $S^n - \{s\}$ is homeomorphic with $\mathbf{R}^n$, so f can be thought of as a map $X \to \mathbf{R}^n$ and, hence, homotopic to a constant map as $\mathbf{R}^n$ is convex.

6.6. $\deg(f \circ g) = \deg(f) * \deg(g)$. Hence $\deg(f \circ g) = \deg(g \circ f)$ and so, by Theorem 6.33, $f \circ g$ and $g \circ f$ are homotopic.

6.7. Yes. No. Yes. No.

Chapter 7

7.3. (1) 1. (2) 0. (3) 0. (4) -2. (5) 0.

7.5. Any negative integer can occur as the Euler number of a connected one-dimensional complex, by taking a triangulation of S^1 and attaching V shapes to 1-simplices, i.e., adding in an extra 0-simplex and two 1-simplices. This reduces the Euler number by 1 so, by iteration, any negative integer can be achieved. However, in a connected complex, 1 is the largest Euler number possible. In a disconnected complex, any positive integer is possible by taking a disjoint collection of 0-simplices.

Chapter 8

8.2. If Y is not path connected, then there are two points y_0, y_1 in Y which cannot be joined by a path. If $f : X \to Y$ is a surjection, then there are points $x_0, x_1 \in X$ such that $f(x_i) = y_i$. Since X is path connected, there is a path $p : [0, 1] \to X$ with $p(0) = x_0$, $p(1) = x_1$. Then $f \circ p$ will be a path from y_0 to y_1 if f is continuous. This cannot happen, by assumption, so f cannot be a continuous surjection $X \to Y$.

8.4. It is multiplication by n.

Chapter 9

9.2. $H_0(\text{square}; \mathbf{Z}/2) = \mathbf{Z}/2$, $H_i(\text{square}; \mathbf{Z}/2) = 0$ for $i > 0$. $H_0(\text{annulus}; \mathbf{Z}/2) = \mathbf{Z}/2$, $H_1(\text{annulus}; \mathbf{Z}/2) = \mathbf{Z}/2$ and $H_i(\text{annulus}; \mathbf{Z}/2) = 0$ for $i > 1$.

Chapter 10

10.1. $H_0(\mathbf{R}^2 - \{0\}) = \mathbf{Z}$, $H_1(\mathbf{R}^2 - \{0\}) = \mathbf{Z}$, $H_i(\mathbf{R}^2 - \{0\}) = 0$ otherwise. $H_0(\mathbf{R}^3 - \{0\}) = \mathbf{Z}$, $H_2(\mathbf{R}^3 - \{0\}) = \mathbf{Z}$, $H_i(\mathbf{R}^3 - \{0\}) = 0$ otherwise.

10.2. $H_0(\mathbf{R}^2 - S^0) = \mathbf{Z}$, $H_1(\mathbf{R}^2 - S^0) = \mathbf{Z} \oplus \mathbf{Z}$, $H_i(\mathbf{R}^2 - S^0) = 0$ otherwise.

10.5. The sequence looks like

$$\cdots 0 \to H_2(X) \to \mathbf{Z} \to \mathbf{Z} \oplus \mathbf{Z} \to H_1(X) \to \mathbf{Z} \to \mathbf{Z} \oplus \mathbf{Z} \to \mathbf{Z} = H_0(X).$$

Working from the right-hand end, we see that the function $H_1(X) \to \mathbf{Z}$ must be 0, so we have $0 \to H_2(X) \to \mathbf{Z} \to \mathbf{Z} \oplus \mathbf{Z} \to H_1(X) \to 0$. Hence $H_2(X)$ is a subgroup of $\mathbf{Z}$, i.e., either 0 or $\mathbf{Z}$. If $H_2(X) = 0$, then $H_1(X) = \mathbf{Z} \oplus \mathbf{Z}/m$ for some integer m, or $H_1(X) = \mathbf{Z}/m \oplus \mathbf{Z}/n$ for some integers m, n. In fact, however, it can be shown geometrically that the map $H_1(U \cap V) \to H_1(U) \oplus H_1(V)$ is 0, so $H_1(X) = \mathbf{Z} \oplus \mathbf{Z}$ and $H_2(X) = \mathbf{Z}$.

10.7. $H_i(\mathbf{Q}) = 0$ for $i > 0$ and $H_0(\mathbf{Q}) = C_0(\mathbf{Q})$ is a free Abelian group with one generator for each rational number.

Bibliography

[1] M.F. Atiyah, *K-Theory*, Benjamin, 1966.

[2] Glen E. Bredon, *Topology and Geometry*, Springer-Verlag, 1993.

[3] S. Eilenberg, N.E. Steenrod, *Foundations of Algebraic Topology*, Princeton Univ. Press, 1954.

[4] William Fulton, *Algebraic Topology: A First Course*, Springer-Verlag, 1995.

[5] Allen Hatcher, *Algebraic Topology*, Cambridge Univ. Press, 2002.

[6] Dale Husemoller, *Fibre Bundles*, third ed., Springer-Verlag, 1994.

[7] William S. Massey, *A Basic Course in Algebraic Topology*, Springer-Verlag, 1991.

Index

Lightning Source UK Ltd.
Milton Keynes UK
UKOW051503261112

202777UK00002B/44/P